ELECTRONIC
CONVERSIONS
SYMBOLS AND FORMULAS
2ND EDITION

ELECTRONIC CONVERSIONS

SYMBOLS AND FORMULAS

2ND EDITION

RUFUS P. TURNER AND STAN GIBILISCO

TAB BOOKS Inc.

Blue Ridge Summit, PA 17294

SECOND EDITION

FIRST PRINTING

Copyright © 1988 by TAB BOOKS Inc.
First Edition Copyright © 1975 by TAB BOOKS Inc.

Printed in the United States of America

Library of Congress Cataloging in Publication Data

Turner, Rufus P.
    Electronic conversions, symbols, and formulas / by Rufus P.
Turner
and Stan Gibilisco. — 2nd ed.
        p.   cm.
    Includes index.
    ISBN 0-8306-0865-6 : ISBN 0-8306-2865-7 (pbk.) :
    1. Electronics—Tables.   2. Electronics—Mathematics.
3. Electronics—Notation.   I. Gibilisco, Stan.   II. Title.
TK7825.T87   1987
621.381′0212—dc19   87-29026
                                                              CIP

Questions regarding the content of this book
should be addressed to:

        Reader Inquiry Branch
        Editorial Department
        TAB BOOKS Inc.
        Blue Ridge Summit, PA 17294

Contents

Introduction

This book is a handy reference source of formulas, tables, symbols, and conversion factors commonly used in electronics. The material is organized by subject for easy access, and should be especially helpful to the electronics technician, engineer, and hobbyist.

The electronics field is made up of numerous specialties. Every possible formula cannot be presented in a book of this size, and, in fact, such a comprehensive volume would be unwieldy and difficult to use. The material in this book has been carefully selected from the personal experience of the authors and also from other electronics professionals to provide the most-needed details.

This second edition includes much information not in the previous edition. The chapter on mathematics has been expanded to provide more formulas and concepts in the general areas of importance in electronics. More detail has been included concerning the nature of complex impedances and the semiconductor information has been updated to keep stride with this most rapidly changing specialty. A complete list of standard schematic symbols has been included. There are more illustrations. Unstinted effort has been exerted to ensure the accuracy of the work, but if errors are found, the authors and publisher would be grateful for reports of them.

It is hoped that this new edition will prove to be valuable in the library of any serious electronics enthusiast.

Chapter 1

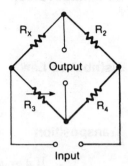

Mathematics

This chapter contains a summary of basic mathematical laws, relationships, symbols, and operations that are frequently encountered in electronics work. The selection of appropriate material has been guided primarily by personal experience and the advice of colleagues. The chapter covers basic arithmetic, algebra, complex numbers, trigonometry, differential and integral calculus, simple probability and statistics, common constants, and mensuration.

1.1 ARITHMETIC AND ALGEBRA

As a simple definition, it can be said that arithmetic involves the manipulation of numbers, while algebra involves the manipulation of symbols. Electronics formulas are written using symbols that represent electrical currents, voltages, etc. The laws of algebra allow us to rearrange these symbols so that we can solve for a particular symbol in terms of the other symbols. The laws of arithmetic then permit us to substitute actual numbers in place of the symbols and so obtain a numerical solution.

The symbols that we use to represent electrical parameters are letters, such as a, b, c, and x, y, z. The manner in which these letter symbols can be manipulated are described in the following topics.

Associative Law

$$(x + y) + z = x + (y + z)$$
$$(xy)z = x(yz)$$

Commutative Law

$$x + y = y + x$$
$$xy = yx$$

Distributive Law

$$x(y + z) = xy + xz$$

Transposition

If $a = b/c$, then $b = ac$, and $c = b/a$.
If $a = bc$, then $b = a/c$, and $c = a/b$.
If $a = b + c$, then $b = a - c$, and $c = a - b$.
If $a = b - c$, then $b = a + c$, and $c = b - a$.
If $a/b = c/d$, then $ad = bc$, $a = bc/d$,
$b = ad/c$, $c = ad/b$, and $d = bc/a$.

Signs

The *sign* of a symbol or number is indicated using a plus (+) or minus (−) sign. These signs indicate whether the number is positive or negative, and whether the number is to be added or subtracted. When two or more signs occur in a mathematical expression, they must obey the following rules.

$$+(+1) = +1 \quad (+1) \times (+1) = +1 \quad (+1)/(+1) = +1$$
$$-(-1) = +1 \quad (-1) \times (-1) = +1 \quad (-1)/(-1) = +1$$
$$+(-1) = -1 \quad (+1) \times (-1) = -1 \quad (+1)/(-1) = -1$$
$$-(+1) = -1 \quad (-1) \times (+1) = -1 \quad (-1)/(+1) = -1$$

Polynomial Products

$$(a + b)(a + b) = a^2 + 2ab + b^2$$
$$(a + b)(a - b) = a^2 - b^2$$
$$(a - b)(a - b) = a^2 - 2ab + b^2$$
$$(a + b)(c - d) = ac + bc - ad - bd$$

Exponents

$$x^a/x^b = x^{(a+b)}$$
$$x^a/x^b = x^{(a-b)} = 1/x^{(b-a)}$$
$$(x^a)^b = x^{ab}$$
$$(x/y)^a = x^a/y^a$$
$$x^{1/a} = \sqrt[a]{x}$$
$$x^{a/b} = \sqrt[b]{x^a}$$
$$x^{-a} = 1/x^a$$
$$x^0 = 1$$
$$x^{-1} = 1/x$$
$$\sqrt{x}/\sqrt{y} = \sqrt{x/y}$$
$$(x^{-a})^{-b} = x^{ab}$$
$$(x^{-a})^b = x^{-ab}$$

Quadratic Equation

If a polynomial is arranged into the form $Ax^2 + Bx + C = 0$, where A, B, and C are constants, then there are two possible solutions which may be obtained using the quadratic equation:

$$x = \frac{-B \pm \sqrt{B^2 - 4AC}}{2A}$$

1.2 MATHEMATICAL SIGNS AND SYMBOLS

.	Radix (base) point
•	Logic multiplication symbol
∞	Infinity
+	Plus, positive, logic OR function
−	Minus, negative
±	Plus or minus, positive or negative
±	Minus or plus, negative or positive
×	Times, logic AND function
÷	Divided by
/	Divided by (expressive of a ratio)
=	Equal to
=	Identical to, is defined by
≅	Approximately equal to, congruent to
≐	Approximately equal to
≠	Not equal to
~	Similar to
<	Less than

≮	Not less than
<<	Much less than
>	Greater than
≯	Not greater than
>>	Much greater than
≤	Equal to or less than
≥	Equal to or greater than
∝	Proportional to, varies directly as
→	Approaches
:	Is to, proportional to
∴	Therefore
#	Number
%	Percent
@	At the rate of; at cost of
ϵ or e	The natural number $\cong$ 2.71828
π	Pi $\cong$ 3.14159 . . .
( )	Parentheses. Used to enclose a common group of terms.
[ ]	Brackets. Used to enclose a common group of terms which includes one or more groups in parentheses.
{ }	Braces. Used to enclose a common group of terms which includes one of more groups in brackets.
∠	Angle
°	Degrees (arc or temperature)
'	Minutes, prime
"	Seconds, double prime
∥	Parallel to
⊥	Perpendicular to
. . .	And beyond, ellipsis

1.3 MATHEMATICAL OPERATIONS

$x+y$	x added to y, x OR y
$x-y$	y subtracted from x
$x \bullet y, x \times y$, or xy	x multiplied by y, x AND y
$x \div y$	x divided by y
x/y or $\frac{x}{y}$	x divided by y
$1/x$	Reciprocal of x
x^n	x raised to the indicated power n
$\sqrt[n]{x}$	Indicated root ($\sqrt[n]{}$) of x
$x{:}y$	x is to y

$\lvert x \rvert$	Absolute value of x, magnitude of x
$X, \vec{X}$, or $\mathbf{X}$	Vector X
$\bar{x}$	Average value of x
$f(x)$ or $F(x)$	Function of x
i	$\sqrt{-1}$
j	Operator, equal to $\sqrt{-1}$
Δx	Increment of x
dx	Differential of x
∂x	Partial differential of x
$\dfrac{\Delta x}{\Delta y}$	Change in x with respect to y
$\dfrac{dx}{dy}$	Derivative of x with respect to y
$\dfrac{d}{dy}(x)$	Derivative of x with respect to y
$D_y x$	Derivative of x with respect to y
$\dfrac{\partial x}{\partial y}$	Partial derivative of x with respect to y
Σ	Summation
Σ_b^a	Summation between limits (from a to b)
Π	Product
Π_b^a	Product between limits (from a to b)
$\int$	Integral
$\int_a^b$	Integral between limits (from a to b)
$\int x\, dy$	Integral of x with respect to y
$\vert_a$	Evaluated at a
$\vert_a^b$	Evaluated between limits (from a to b)

1.4 MATHEMATICAL FUNCTIONS

In mathematics, a function describes the way in which one quantity varies with respect to other quantities. Thus, if y varies in proportion to x^2, we say that y *is a function of x*. We write this relationship as $y = f(x)$, where $f(\ )$ means *function of*. Since $y = x^2$ in this example, we can fully describe the function $f(x)$ by writing $f(x) = x^2$. Most often, however, we either do not know the exact relationship between the quantities or we do not wish repeatedly to describe the relationship in detail.

There are many common functions occurring in mathematics which are represented by short abbreviations, such as *cos* and *log*. It is traditionally acceptable to write such functions without using parentheses; for example, to write cos A instead of cos(A). However, parentheses must be used whenever their omission would cause confusion; for example, cos A + B might be interpreted as either cos (A) + B or cos $(A + B)$.

Trigonometric and Hyperbolic Functions

sin x	Sine of angle x
sinh x	Hyperbolic sine of angle x
cos x	Cosine of angle x
cosh x	Hyperbolic cosine of angle x
tan x	Tangent of angle x
tanh x	Hyperbolic tangent of angle x
cot x	Cotangent of angle x
coth x	Hyperbolic cotangent of angle x
sec x	Secant of angle x
sech x	Hyperbolic secant of angle x
csc x	Cosecant of angle x
cosech x	Hyperbolic cosecant of angle x
covers$x$	Coversed sine (the versed sine of the complement of angle x)
versin x	Versed sine (one minus cosine of angle x)

Inverse Trigonometric Functions

arc	Prefix indicating the inverse function, as arcsin $x = \sin^{-1}x$
sin ^{-1}x	Angle whose sine is x
sinh^{-1}x	Angle whose hyperbolic sine is x
cos^{-1}x	Angle whose cosine is x
cosh^{-1}x	Angle whose hyperbolic cosine is x
tan^{-1}x	Angle whose tangent is x
tanh^{-1}x	Angle whose hyperbolic tangent is x
cot^{-1}x	Angle whose cotangent is x
coth^{-1}x	Angle whose hyperbolic cotangent is x
sec^{-1}x	Angle whose secant is x
sech^{-1}x	Angle whose hyperbolic secant is x
csc^{-1}x	Angle whose cosecant is x
cosech^{-1}x	Angle whose hyperbolic cosecant is x

Logarithms

If $a^x = b$, then $\log_a b = x$, where a is said to be the *base* of the logarithm. For *natural* or *napierian* logarithms, the base is $\epsilon \cong 2.71828$. For *common* logarithms, the base is 10.

In common logarithms, the base is normally omitted, and the logarithm is indicated as $\log x$. To avoid confusion, it is common practice to use the symbol ln to indicate the natural logarithm.

The following relationships are useful when working with logarithms:

$$
\begin{aligned}
\log_{10} b &= \log_\epsilon b / \log_\epsilon 10 \cong 0.434 \log_\epsilon b \\
\log_\epsilon b &= (\log_{10} b)/(\log_{10}\epsilon) \cong 2.303 \log_{10} b \\
\log ab &= \log a + \log b \\
\log(a/b) &= -\log(b/a) = \log a - \log b \\
\log a^x &= x \log a \\
\log(1/a) &= -\log a \\
\log(\sqrt[n]{a}) &= (\log a)/n \\
\log_\epsilon \epsilon^x &= x \\
\log_{10} 10^x &= x
\end{aligned}
$$

Antilogarithms and Cologarithms

The antilogarithm (antilog or $\log^{-1}$) is the number whose logarithm is x. If $\log b = x$, then $b = \text{antilog } x = \log^{-1} x$.

$$
\begin{aligned}
\text{antilog}_a x &= a^x \\
\text{antilog}_\epsilon x &= e^x \\
\text{antilog}_{10} x &= 10^x
\end{aligned}
$$

The cologarithm (colog) is the logarithm of the reciprocal of a number.

$$
\text{colog } a = \log \frac{1}{a} = -\log a
$$

1.5 FACTORIALS

The factorial (!) of a number is the product of all the positive integers in that number. Thus, $5! = 5 \times 4 \times 3 \times 2 \times 1 = 120$. Factorials between 1 and 10 are given below:

$$
\begin{aligned}
1! &= 1 & 6! &= 720 \\
2! &= 2 & 7! &= 5040 \\
3! &= 6 & 8! &= 40{,}320
\end{aligned}
$$

$$4! = 24 \qquad 9! = 362{,}880$$
$$5! = 120 \qquad 10! = 3{,}628{,}800$$

Stirling's Formula

$$n! \cong (2\pi)^{1/2} n^{(n+1/2)} e^{-n}$$

1.6 COMPLEX NUMBERS

A complex number consists of a real number and an imaginary number. Complex numbers are very important as a tool to enable us to work out complex mathematical problems in which the expression $\sqrt{-1}$ occurs. From a practical standpoint, we cannot take the square root of a negative number, so that any solution we do arrive at has to be purely imaginary.

In mathematics, it is traditional to use the letter i to represent the square root of minus one. However, in electronics the letter i is generally used to represent current. Thus in electronics it has become traditional to use the letter j instead of i.

$$
\begin{aligned}
j^0 &= 1 \\
j &= \sqrt{-1} \\
j^2 &= -1 \\
j^3 &= -\sqrt{-1} = -j \\
j^4 &= +1 \\
j^{-1} &= 1/j = -\sqrt{-1} = -j
\end{aligned}
$$

Mathematical Operations Involving Complex Numbers

$$
\begin{aligned}
(a + jb)^2 &= (a^2 - b^2) + 2jab \\
(a + jb)(a - jb) &= a^2 + b \\
(a + jb) + (c + jd) &= (a + c) + j(b + d) \\
(a + jb) - (c + jd) &= (a - c) + j(b - d) \\
(a + jb)(c + jd) &= (ac - bd) + j(bc + ad) \\
(a + jb)(c - jd) &= (ac + bd) + j(bc - ad)
\end{aligned}
$$

$$
(a + jb)/(c + jd) = \frac{(a + jb)(c - jd)}{(c + jd)(c - jd)}
$$

$$
= \frac{(ac + bd) + j(bc - ad)}{c^2 + d^2}
$$

1.7 CIRCULAR TRIGONOMETRY

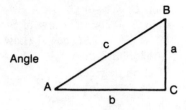

Fig. 1-1. Definitions of trigonometric functions.

Sides of Right Triangle

$$a = \sqrt{c^2 - b^2}$$
$$b = \sqrt{c^2 - a^2}$$
$$c = \sqrt{a^2 + b^2}$$

Angles of Right Triangle

$$C = 90°$$
$$A = 90° - B$$
$$B = 90° - A$$

Functions of Angle

$$\sin A = a/c$$
$$\cos A = b/c$$
$$\tan A = a/b$$
$$\cot A = b/a$$
$$\sec A = c/b$$
$$\csc A = c/a$$

$$\sin(a \pm b) = \sin a \cos b \pm \cos a \sin b$$

$$\cos(a \pm b) = \cos a \cos b \mp \sin a \sin b$$

$$\tan(a \pm b) = \frac{\tan a \pm \tan b}{1 \mp \tan a \tan b}$$

Useful Relationships

$$\sin -A = -\sin A$$
$$\cos -A = \cos A$$
$$\tan -A = -\tan A$$
$$\cot A = 1/\tan A$$
$$\sec A = 1/\cos A$$
$$\csc A = 1/\sin A$$
$$\sin^2 A + \cos^2 A = 1$$
$$\cos A = \sqrt{1 - \sin^2 A}$$

$$\sin A = \sqrt{1 - \cos^2 A}$$
$$\tan A = \sin A/\cos A$$
$$\tan A = b/\sin B = c/\sin C$$
$$a/\sin A = b/\sin B = c/\sin C \text{ (Law of Sines)}$$
$$A^2 = B^2 + C^2 - 2BC \cos A \text{ (Law of Cosines)}$$
$$1 \text{ degree} = 0.0174533 \text{ radian}$$
$$1 \text{ radian} = 57.2958 \text{ degrees}$$

When A is in radians,

$$\sin A = A - (A^3/3!) + (A^5/5!) - (A^7/7!) + \ldots$$
$$\cos A = 1 - (A^2/2!) + (A^4/4!) - (A^6/6!) + \ldots$$

Signs of Functions in Different Quadrants

Quadrant	I	II	III	IV
A, degrees	0 – 90	90 – 180	180 – 270	270 – 360
A, radians	0 – $\pi/2$	$\pi/2 - \pi$	$\pi - 3\pi/2$	$3\pi/2 - 2\pi$
sin A or csc A	+	+	−	−
cos A or sec A	+	−	−	+
tan A or cot A	+	−	+	−

1.8 HYPERBOLIC TRIGONOMETRY

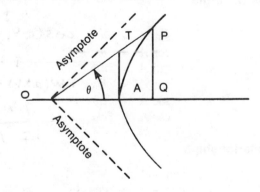

Fig. 1-2. Definitions of hyperbolic functions.

Functions of Angle

$$\sinh \theta = PQ = (\epsilon^\theta - \epsilon^{-\theta})/2$$
$$\cosh \theta = OQ = (\epsilon^\theta + \epsilon^{-\theta})/2$$
$$\tanh \theta = BA = \sinh\theta/\cosh\theta$$
$$\coth \theta = 1/BA = 1/\tanh\theta$$

$$\text{sech} \quad \theta \;=\; 1/OQ \;=\; 1/\cosh \theta$$
$$\text{cosech} \quad \theta \;=\; 1/PQ \;=\; 1/\sinh \theta$$

Useful Relationships

$$\cosh^2 \theta - \sinh^2 \theta = 1$$
$$\sinh \theta = \sqrt{\cosh^2 \theta - 1}$$
$$\cosh \theta = \sqrt{\sinh^2 \theta + 1}$$
$$\tanh \theta = \sinh \theta / \sqrt{\sinh^2 \theta + 1}$$

1.9 ELEMENTARY DIFFERENTIAL CALCULUS

The *derivative* of a variable x with another variable t is written dx/dt. A derivative may be described with respect to any other variable in a relation or function.

A full description of differential calculus is beyond the scope of this book. However, in general, a derivative is the instantaneous rate of change of the dependent variable in a function, with respect to the independent variable. We might illustrate an example by showing the acceleration of an automobile as it goes from 0 to 50 miles per hour in 50 seconds.

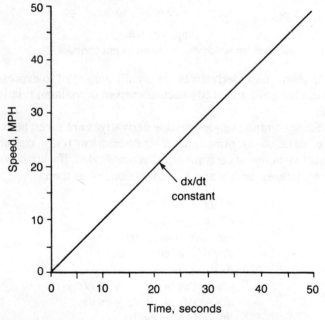

Fig. 1-3. A function for which the derivative is a constant.

Assuming the rate of speed increase is constant, the car is increasing speed at 1 mph/sec during the whole interval. But, the rate of increase need not necessarily be constant, and therefore the derivative of the speed x relative to time t depends on t.

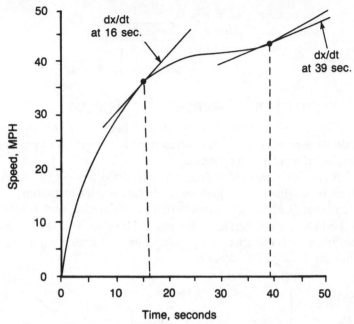

Fig. 1-4. A function for which the derivative is not constant.

In electronics, derivatives are usually employed to express the rate of change of a quantity such as current or voltage, relative to time.

Some common single-variable derivatives are given here. For more detailed information, a mathematics text, containing exhaustive tables of derivatives, is recommended. If a is a constant, n is an integer, and u and v are functions of x, then:

$$da/dx = 0$$
$$d(ax)/dx = a$$
$$d(au)/dx = a(du/dx)$$
$$d(x^n)/dx = n(x^{n-1})$$
$$dx/dx = 1$$
$$d(u+v)/dx = du/dx + dv/dx$$
$$d(u-v)/dx = du/dx - dv/dx$$
$$d(nu)/dx = n(du/dx)$$

$$d(uv)/dx = u(dv/dx) + v(du/dx)$$
$$d(u^n)/dx = n(u^{n-1})(du/dx)$$

$$d(u/v)/dx = \frac{v(du/dx) - u(dv/dx)}{v^2}$$

$$d(x^2)/dx = 2x$$
$$d(x^3)/dx = 3x^2$$
$$d(x^4)/dx = 4x^3$$
$$d(x^{-1})dx = \ln x$$
$$d(\ln x)/dx = x^{-1}$$
$$d(\log_{10} x)/dx = 0.343/x \text{ (approx.)}$$
$$d(e^x)/dx = e^x$$
$$d(e^u)/dx = e^u(du/dx)$$
$$d(e^{ax})/dx = a(e^x)$$
$$d(\ln u)/dx = u^{-1}(du/dx)$$
$$d(n^u)/dx = n^u \ln n(du/dx)$$
$$d(u^v)/dx = v(u^{v-1})(du/dx) + (\ln u)\, u^v\, (dv/dx)$$
$$d(e^u)/dx = e^u(du/dx)$$
$$d(\sin x)/dx = \cos x$$
$$d(\cos x)/dx = -\sin x$$
$$d(\tan x)/dx = \sec^2 x$$
$$d(\cot x)/dx = \csc^2 x$$
$$d(\sec x)/dx = \sec x \tan x$$
$$d(\csc x)/dx = -(\csc x \cot x)$$
$$d(\sin u)/dx = \cos u\, (du/dx)$$
$$d(\cos u)/dx = \sin u\, (du/dx)$$
$$d(\tan u)/dx = \sec^2 u\, (du/dx)$$
$$d(\cot u)/dx = \csc^2 u\, (du/dx)$$
$$d(\sec u)/dx = \sec u \tan u\, (du/dx)$$
$$d(\csc u)/dx = -(\csc u \cot u)\, (du/dx)$$
$$d(\arcsin x)/dx = (1-x^2)^{-1/2}$$
$$d(\arccos x)/dx = -(1-x^2)^{-1/2}$$
$$d(\arctan x)/dx = 1/(1+x^2)$$
$$d(\text{arccot } x)/dx = -1/(1+x^2)$$
$$d(\text{arcsec } x)/dx = 1/(x\sqrt{x^2-1})$$
$$d(\text{arccsc } x)/dx = -1/(x\sqrt{x^2-1})$$
$$d(\arcsin u)/dx = (1-u^2)^{-1/2}\, (du/dx)$$
$$d(\arccos u)/dx = -(1-u^2)^{-1/2}(du/dx)$$
$$d(\arctan u)/dx = (du/dx)/(1+u^2)$$
$$d(\text{arccot } u)/dx = -(du/dx)/(1+u^2)$$
$$d(\text{arcsec } u)/dx = (du/dx)/(u\sqrt{u^2-1})$$

$$d(\text{arccsc } u)/dx = -(du/dx)/(u\sqrt{u^2-1})$$
$$d(\sinh x)/dx = \cosh x$$
$$d(\cosh x)/dx = \sinh x$$
$$d(\tanh x)/dx = \text{sech}^2 x$$
$$d(\sinh u)/dx = \cosh u \ (du/dx)$$
$$d(\cosh u)/dx = \sinh u \ (du/dx)$$
$$d(\tanh u)/dx = \text{sech}^2 u \ (du/dx)$$

1.10 ELEMENTARY INTEGRAL CALCULUS

The *integral* of a function in one variable may broadly be defined as an expression of the area under a curve. Integrals are expressed either in indefinite form or definite form. The indefinite integral is the area under a curve from 0 to any value on the abscissa. This value is not specified.

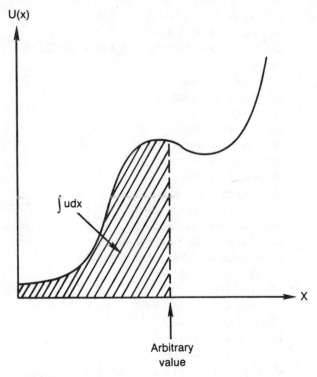

Fig. 1-5. Indefinite integral.

The definite integral is the area under the curve between two certain abscissa values, say x_1 and x_2.

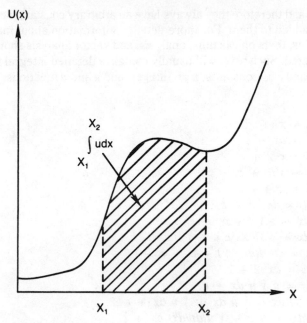

Fig. 1-6. Definite integral.

The indefinite integral of a function u is written

$$\int u \, dx,$$

where x is the independent variable. The definite integral between values x_1 and x_2 is written

$$\int_{x_1}^{x_2} u \, dx$$

and is equal to the value of the indefinite integral evaluated at x_2, minus the value of the indefinite integral evaluated at x_1:

$$\int_{x_1}^{x_2} u \, dx = \int u \, dx \mid_{x_1} - \int u \, dx \mid_{x_2}$$

In electronics, integrals are applied in a wide variety of situations, and can become quite complicated in nature. Some common single-variable integrals are given here. All are in indefinite

15

form, and therefore they always have an arbitrary constant, denoted by c, added to them. For more detailed information about integral calculus, texts on calculus, analysis, and vector analysis should be consulted. Such texts will usually contain a detailed integral table. Let a and c be constants, n an integer, and u and v functions. Then

$$\int 0 \, dx = c$$
$$\int 1 \, dx = x + c$$
$$\int a \, dx = ax + c$$
$$\int x \, dx = x^2/2 + c$$
$$\int x^2 \, dx = x^3/3 + c$$
$$\int x^3 \, dx = x^4/4 + c$$
$$\int x^{-1} \, dx = ln \, |x| + c$$
$$\int x^{-2} \, dx = -1/x + c$$
$$\int x^{-3} \, dx = -1/(2x^2) + c$$
$$\int x^n \, dx = x^{n+1}/(n+1)) + c, \, n \neq -1$$
$$\int ax \, dx = ax^2/2 + c$$
$$\int au \, dx = a \int u \, dx + c$$
$$\int (u + v) \, dx = \int u \, dx + \int v \, dx + c$$
$$\int v(du/dx) = u \, v - \int u(dv/dx) \, dx + c$$
$$\int e^x \, dx = e^x + c$$
$$\int e^{ax} \, dx = e^x/a + c$$
$$\int ln \, x \, dx = x \, ln \, x - x + c$$
$$\int \log_{10} x \, dx = 0.343 \, (ln \, x - 1) + c \, \text{(approx.)}$$
$$\int \sin x \, dx = -\cos x + c$$
$$\int \cos x \, dx = \sin x + c$$
$$\int \tan x \, dx = -ln \, |\cos x|$$
$$\int \cot x \, dx = ln \, |\sin x|$$
$$\int \sec x \, dx = ln \, |\sec x + \tan x|$$
$$\int \csc x \, dx = ln \, |\tan (x/2)|$$
$$\int \sec^2 x \, dx = \tan x + c$$
$$\int \int - \csc^2 x \, dx = \cot x + c$$
$$\int (\sec x \tan x) \, dx = \sec x + c$$
$$\int \sinh x \, dx = \cosh x + c$$
$$\int \cosh x \, dx = \sinh x + c$$
$$\int \tanh x \, dx = ln \, | \cosh x |$$
$$\int \coth x \, dx = ln \, |\sinh x |$$
$$\int \operatorname{sech} x \, dx = 2 \arctan e^x$$
$$\int \operatorname{csch} x \, dx = ln \, |\tanh (x/2) |$$

In general, the indefinite integral of a function is the opposite of the derivative, plus a constant. In many cases integrals can be found by means of a derivative table, by simply working backwards and adding a constant.

1.11 SERIES

A *series* is the sum of a sequence of numbers. Of particular interest are the infinite series, that is, series with an infinite number of addends. The sum may or may not be finite. For example, the sum of the series $S = 1 + 2 + 3 + 4 + \ldots$ is infinite, while the sum $S = 1 + \frac{1}{2} + \frac{1}{4} + \frac{1}{8} + \frac{1}{16} + \frac{1}{32} + \ldots = 2$. Series whose sums are not finite are said to be *divergent*. Those with finite sums are said to be *convergent*.

Series are of special importance when they result in functions of a variable. In electronics, infinite series are useful especially in waveform analysis, since certain waveshapes are the result of a combination of infinitely many others.

Series are generally denoted in either of two ways: the sequential method, as shown above, and the summation method. Let S be the sum of a sequence of n individual numbers denoted by x_k, where k may be any positive integer. The sequential method yields the expression

$$S = x_1 + x_2 + x_3 + \ldots + x_k$$

If the series is infinite, then we write

$$S = x_1 + x_2 + x_3 \ldots$$

The summation method makes use of the uppercase Greek letter sigma (Σ) to write a finite series as

$$S = \sum_{k=1}^{n} x_k,$$

where n is the largest value of k. For an infinite series,

$$S = \sum_{k=1}^{\infty} x_k$$

The theory of infinite series is quite involved, and a complete discussion is beyond the scope of this book. Certain infinite series have well-known functional sums. The more common are shown below.

$$1 + \tfrac{1}{2} + \tfrac{1}{4} + \tfrac{1}{8} + \tfrac{1}{16} + \ldots = 2$$

$$1 - \tfrac{1}{2} + \tfrac{1}{4} - \tfrac{1}{4} + \tfrac{1}{5} + \ldots = \ln 2 = 0.693147$$

$$1 + \tfrac{1}{2}^2 + \tfrac{1}{3}^2 + \tfrac{1}{4}^2 + \tfrac{1}{5}^2 + \ldots = \pi^2/6 = 1.64493$$

$$1 - \tfrac{1}{2}^2 + \tfrac{1}{3}^2 - \tfrac{1}{4}^2 + \tfrac{1}{5}^2 + \ldots = \pi^2/12 = 0.822467$$

$$1 - x + x^2 - x^3 + x^4 - x^5 + \ldots = 1/(1 + x) \text{ for } -1 < x < 1$$

$$1 + x + x^2/2! + x^3/3! + x^4/4! + x^5/5! + \ldots = e^x$$

$$x - x^2/2 + x^3/3 - x^4/4 + x^5/5 + \ldots = \ln(1 + x)$$

for $-1 < x <$

$$2\left(\frac{x-1}{x+1}\right) + \frac{2}{3}\left(\frac{x-1}{x+1}\right)^3 + \frac{2}{5}\left(\frac{x-1}{x+1}\right)^5 + \frac{2}{7}\left(\frac{x-1}{x+1}\right)^7 +$$

$$\frac{2}{9}\left(\frac{x-1}{x+1}\right)^9 + \ldots = \ln x$$

$$x - x^3/3! + x^5/5! - x^7/7! + x^9/9! + \ldots = \sin x$$

$$1 - x^2/2! + x^4/4! - x^6/6! + x^8/8! + \ldots = \cos x$$

$$x + x^3/3! + x^5/5! + x^7/7! + x^9/9! + \ldots = \sinh x$$

$$1 + x^2/2! + x^4/4! + x^6/6! + x^8/8! + \ldots = \cosh x$$

Fig. 1-7. Examples of frequently encountered series.

Fourier Series

$$f(x) = (A_0/2) + \sum_{k=1}^{\infty} [A_k \cos(kx) + B_k \sin(kx)]$$

where A_0, A_k and B_k are constants. Any periodic function can be expressed in terms of a Fourier series. The more terms evaluated, the more accurate the representation of the function.

Power Series

$$S = a_0 + a_1 x + a_2 x^2 + \ldots = \sum_{k=0}^{n} a_k x^k$$

where x is a variable that may attain any real-number value.

A power series may or may not have a finite sum. For certain values of the a factors, the power series will have a finite sum no matter what the value of x. For other values of the a factors, the series will converge for some values of x but not for others. For some values of the a factors, the series will have a finite sum only if $x = 0$; this sum will invariably be 0.

Power series are useful in the design of bandpass, lowpass, highpass, and band-rejection filters.

Taylor Series

Let $T(x)$ represent the Taylor function of a variable x, and let $f(x)$ be some function of x. Then

$$T(x) = f(a) + f'(a)(x-a) + f^{11}(a)(x-a)^2/2! + f'''(a)(x-a)^3/3! + \ldots$$

The larger the number of terms evaluated, the closer the series approximates the Taylor function. Taylor series are useful in the design of selective filters and also in waveform analysis.

1.12 PROBABILITY AND STATISTICS

This section covers elementary probability and statistics. Such basics as computing simple probabilities and averages are provided in the following topics.

Mathematical Probability

If an event can succeed in any one of s equally likely ways and can fail in any one of f equally likely ways, then:

Probability of success, $P_s = s/(s + f)$
Probability of failure, $P_f = f/(s + f)$

Permutations

The number of permutations of n things taken r at a time is

$$_nP_r = n!/(n - r)!$$

Combinations

The number of combinations of n things taken r at a time is

$$_nC_r = \frac{n!}{r!(n - r)!}$$

Average Values

Arithmetic mean:

$$\overline{x} = \frac{x_1 + x_2 + \ldots + x_n}{n}$$

Geometric mean:

$$\overline{x} = \sqrt[n]{x_1 x_2 x_3 \ldots x_n}$$

Harmonic mean:

$$\overline{x} = \frac{1}{1/n(1/x1 + 1/x2 + 1/x3 + \ldots + 1/xn)}$$

Root-mean-square (rms) or standard deviation (σ):

$$x_{rms} = \sigma = \sqrt{\frac{x_1^2 + x_2^2 + x_3^2 + \ldots + x_n^2}{n}}$$

1.13 COMMON CONSTANTS

$$
\begin{aligned}
\epsilon &= 2.71828 \\
\epsilon^2 &= 7.38906 \\
1/\epsilon &= 0.367879 \\
\pi &= 3.14159 \\
\pi^2 &= 9.86960 \\
2\pi &= 6.28318 \\
4\pi &= 12.5664 \\
\sqrt{2} &= 1.41421 \\
1/\sqrt{2} &= 0.707107 \\
\sqrt{3} &= 1.73205 \\
\pi/2 &= 1.57080 \\
1/\pi &= 0.31831 \\
\tfrac{1}{2}\pi &= 0.15915
\end{aligned}
$$

Boltzmann's constant, $k = 1.380 \times 10^{-23}$ J/°K
Euler's constant, $\gamma = 0.577216$
Planck's constant, $h = 6.6236 \times 10^{-36}$ J-sec
Speed of light (in vacuum), $c = 186,000$ mi/sec $= 300,000$ km/sec
Speed of sound (in air at sea level), $v = 1129$ ft/sec

1.14 MENSURATION AND MISCELLANEOUS

In the following formulas,

A = area
B = length of base
C = circumference
D = diagonal or diameter
H = height
L = length
P = perimeter
R = radius
S = length of a side
V = volume
π = constant $\cong$ 3.14159

Areas

Circle: $A = \pi R^2 = \pi D^2/4$
Ellipse: $A = 0.7854 \times$ major axis $\times$ minor axis
Parabola: $A = \frac{2}{3} BH$
Sphere (surface): $A = 4\pi R^2$
Square or rectangle: $A = LW$
Triangle: $A = \frac{1}{2} BH$

Volumes

Cone: $V = \frac{1}{3} AH$
Cube: $V = LWH$
Cylinder (right circular): $V = 2\pi RH$
Prism: $V = AH$
Pyramid: $V = \frac{1}{3} AH$
Sphere: $V = 4\pi R^3/3$

Perimeter of Square or Rectangle

$P = S_1 + S_2 + S_3 + S_4$

Circumference of Circle

$C = 2\pi R$

Diagonal of Square

$D = 1.414S$

1.15 SCIENTIFIC NOTATION

It is not always convenient to express numbers of very large or very small absolute value. For example, the number 10 quadrillion is written 10,000,000,000,000,000, and the number 0.1 quadrillionth is written 0.0000000000000001 in conventional form. Extreme values are so unwieldy as to appear ridiculous in the usual decimal form. In such cases, power-of-ten form is preferred.

The standard form involves the use of a number greater than or equal to 1, but less than 10; that is, from 1.0000 . . . to 9.9999 . . . , followed by a multiplication sign and a power of 10.

Powers of 10

10^{-15} = 1 quadrillionth
10^{-12} = 1 trillionth
10^{-9} = 1 billionth
10^{-6} = 1 millionth
10^{-3} = 1 thousandth
10^{0} = 1
10^{3} = 1 thousand
10^{6} = 1 million
10^{9} = 1 billion
10^{12} = 1 trillion
10^{15} = 1 quadrillion

In this form, we write 1,000 as 1×10^3; 10,000 would be written 1×10^4; 0.001 is written as 1×10^{-3}. Other examples: $3,500 = 3.5 \times 10^3$; $0.0002322 = 2.322 \times 10^{-4}$. Extreme numbers become much simpler to write. An example of this is three googol, 3×10^{100}. Sometimes with numbers in which the whole-number part is not followed by a decimal fraction, we may write a decimal point followed by the number of zeroes required by the degree of accuracy; for example, 3.00×10^7.

1.16 DEGREE OF ACCURACY

The degree of accuracy is indicated by the number of digits in the decimal part of a scientific-notation expression. The value of 25,440, for example, might be written 2.544×10^4. This expression has four significant digits. But 25,441, written 2.5441×10^4, has five significant digits. A very exact expression such as 25,441.5656 is written 2.5515656×10^4 and is accurate to eight digits.

The accuracy of an expression in electronics is limited by our ability to measure it—and this is rarely more than ten significant figures. However it may be only one or two.

1.17 ROUNDING

In making calculations we may have values that vary according to the number of significant figures. For example, we might measure a current of 1.456 A at 117 volts. The power drawn in this case can be expressed to only three significant digits: $P = 1.456 \times 117 = 1.456 \times 10^0 \times 1.17 \times 10^2$. This is normally done by multiplying out the expression and then rounding the result according to the fewest significant figures in any expression. Thus we obtain $1.456 \times 1.17 \times 10^0 \times 10^2 = 1.70352 \times 10^2$, but the decimal part must be rounded to three significant digits. This gives $P = 70$ W as the power drawn. We cannot justify more than the least number of significant digits in the final result.

The basic rule of rounding is that a final digit of 0 to 4 is rounded to zero, and a final digit of 5 to 9 is rounded to 0 and then the digit immediately to the left is changed to 1. This process is repeated until the required number of significant digits is reached. An example appears below, in which 2.47878922402 is rounded to three significant digits.

Rounding of Significant Digits
2.47878922402
2.4787892240
2.478789224
2.47878922
2.4787892
2.478789
2.47879
2.4788
2.479
2.48

1.18 CARTESIAN COORDINATE SYSTEM

Values are not always expressed as single-dimensional quantities. Simple numbers are called scalars. Such numbers are denoted by the simple numerals and variables of arithmetic and algebra. They can be shown geometrically as a simple, linear number line. This is the *Real Number Line*. This line is infinitely long, including zero and all of the positive and negative real numbers.

Fig. 1-8. A simple number line.

The real number line can be nonlinear, such as in a logarithmic scale. This kind of scale is designed according to power-of-ten format, and can include all positive or negative real numbers except zero.

Two-dimensional quantities can be expressed by arranging two number lines perpendicular to each other. This scheme is called the *Cartesian coordinate system*. It is in theory a geometric plane. Several two-dimensional values are shown. The variables are denoted in mathematics as x (the horizontal axis or abscissa) and y (the vertical axis or ordinate). In electronics these values may take on other meanings, such as time and temperature.

In general form a point on the Cartesian plane is denoted by an ordered number pair (x,y). Thus the point $x = 3$ and $y = -5$ is written $(3, -5)$. This point is shown in the illustration.

Distance Between Two Points

$d = \sqrt{(y_2 - y_1)^2 + (x_2 - x_1)^2}$, where the points are (x_1, y_1) and (x_2, y_2).

Distance Between Origin (0,0) and Point (x,y)

$$d = \sqrt{y^2 - x^2}$$

This distance is also called the radius, or the length of a vector $\overrightarrow{\mathbf{x,y}}$ beginning at the origin.

1.19 LOGARITHMIC AND SEMILOGARITHMIC COORDINATES

A variation of the Cartesian coordinate system is found in the logarithmic and semilogarithmic schemes. In the logarithmic or log-log system, both axes are logarithmic rather than linear. This results

24

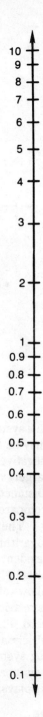

Fig. 1-9. Logarithmic number line.

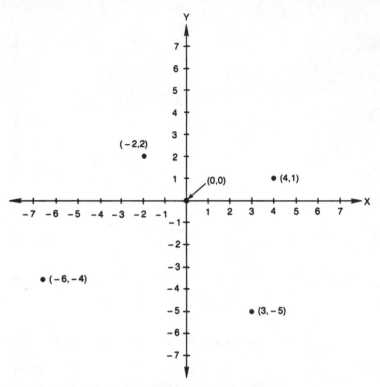

Fig. 1-10. Cartesian plane showing origin and four points.

in a system where zero is not shown. The abscissa and the ordinate may be both positive, both negative, or one positive and the other negative. These systems, each shown below at A, B, C and D, allow larger values to be displayed with respect to smaller values. Distances are distorted, as are the actual graphical displays, as compared with the linear Cartesian plane.

The semilogarithmic scheme has one linear axis and one logarithmic axis. The linear axis is usually the abscissa, or independent variable, and the logarithmic axis the dependent variable, although sometimes it is the other way around. The linear axis allows all of the real numbers to be shown, but the logarithmic axis is restricted to numbers either positive or negative only, excluding zero. There are four basic schemes for the semilogarithmic linear abscissa graph system, shown at A, B, C and D. As with the log-log system, distances are distorted by comparison with the conventional linear graph. The same is true of the actual graphical displays.

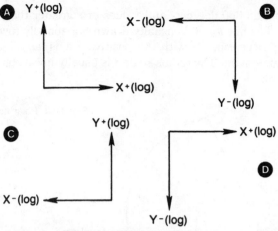

Fig. 1-11. Log-log coordinate systems. At A, both *x* and *y* are positive; at B, *x* negative and *y* negative; at C, *x* negative and *y* positive; at D, *x* positive and *y* negative.

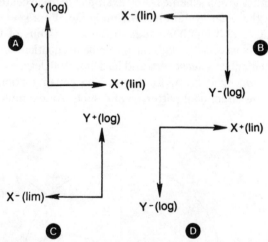

Fig. 1-12. Semi-log coordinates. At A, both x and y are positive; at B, both x and y are negative; at C, x negative and y positive; at D, x positive and y negative.

1.20 POLAR COORDINATES

A different method of expressing an ordered number pair is by means of an angle and a distance from the origin. Generally the angle is called θ (theta) and the distance r (radius). Either θ or r can be positive or negative. However, all the points on a geometric plane can be represented by angles θ greater than or equal to zero but less than 360 degrees, and values of r greater than or equal to zero. The line $\theta = 0$ is usually drawn horizontally towards the right, and corresponds with the positive x axis (abscissa) of the Cartesian system. The radius scale r is usually linear, but may be logarithmic.

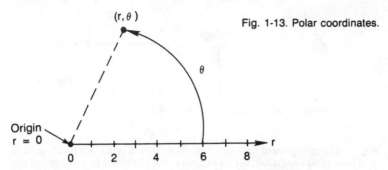

Fig. 1-13. Polar coordinates.

Some expressions are much easier to denote in the polar system than in other graph schemes. For example, a circle centered at the origin in the polar system has a simple equation: $r = x$, where x is a positive real number equal to the unit radius of the circle. Straight rays passing through the origin have equations $\theta = y$, where y is a real number at least zero and less than 360 degrees. The polar system is especially useful for showing antenna radiation patterns, microphone directional patterns, and such. An example is shown below.

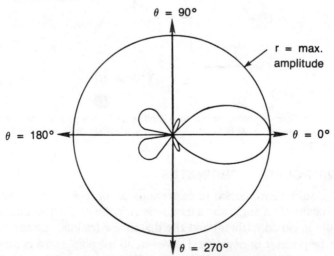

Fig. 1-14. Example of a directional pattern in polar coordinates.

28

Conversion from Linear Cartesian
Coordinates (x,y) to Polar (r,θ)

$$r = \sqrt{x^2 + y^2}$$
$$\theta = \arctan(y/x)$$

Conversion from Polar
Coordinates to Cartesian Coordinates

$$x = r\cos\theta$$
$$y = r\sin\theta$$

1.21 COORDINATES IN THREE DIMENSIONS

The most common coordinate system used in three dimensions is a Cartesian linear system. There are three axes, usually called the x, y, and z axes. Each axis is perpendicular to the other two.

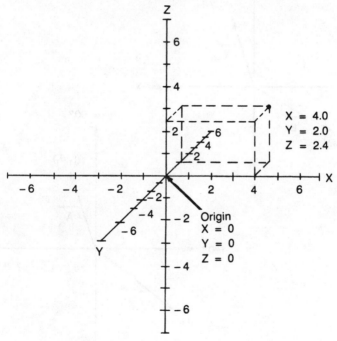

Fig. 1-15. Cartesian coordinates in three dimension.

The most often-used scheme consists of three linear axes. The location of any point in space can be uniquely determined by an ordered triple (x,y,z). This system is hard to visualize without the aid of computer graphics. Plots in three dimensions may include lines, curves, helices, spirals, cones, spheres, and a large variety of other objects. Theoretically, any object imaginable can be expressed as an equation in three dimensions.

Three-dimensional coordinates are generally used to show the complete spatial directional patterns for antennas, microphones, and the like.

1.22 VECTORS

A vector is a quantity that has magnitude and direction. It can be expressed in one, two, or three dimensions. Examples of vectors include force, velocity, acceleration, current, and the propagation of an electromagnetic field.

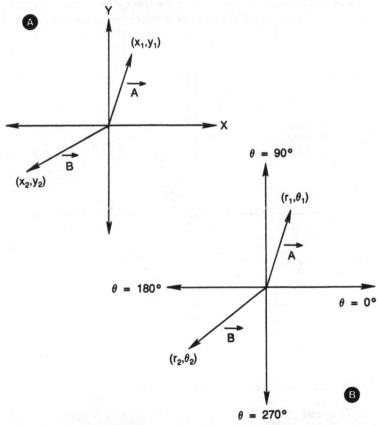

Fig. 1-16. At A, examples of vectors in Cartesian coordinates. At B, the same vectors in polar coordinates.

Vectors in a plane can be represented either by Cartesian coordinates (x,y), or by polar coordinates (r,θ). These are ordered number pairs corresponding to points in the plane. The vector is considered to begin at the origin and end at the point. The magnitude of a vector in Cartesian coordinates is given by

$$A = \sqrt{x^2 + y^2},$$

where A represents the magnitude of vector $\vec{A}$. In polar coordinates, the magnitude is simply the value of r.

The direction of a vector is simple to express in the polar system. It is simply the value of θ. The direction in Cartesian coordinates is the arctangent of y/x. Thus if we let A be the magnitude and d the direction for a given vector $\vec{A} = (\vec{x,y}) = (\vec{r,\theta})$:

Magnitude of Vector in Polar Coordinates

$$A = r$$

Magnitude of Vector in Cartesian Coordinates

$$A = \sqrt{x^2 + y^2}$$

Direction of Vector in Polar Coordinates

$$d = \theta$$

Direction of Vector in Cartesian Coordinates

$$d = \arctan(y/x)$$

Vectors can be added by the familiar parallelogram method. In Cartesian coordinates, this is easy to show geometrically. If vector $\vec{A} = (\vec{x_1,y_1})$ and $\vec{B} = (\vec{x_2, y_2})$, we have

Sum of Vectors in Cartesian Coordinates

$$\vec{A} + \vec{B} = (\vec{x_1 + x_2, y_1 + y_2})$$

That is, the x components are added and the y components are added. To subtract two vectors, we take the negative vector and add it to the other vector.

Difference of Vectors in Cartesian Coordinates

$$\vec{A} - \vec{B} = \vec{A} + (-\vec{B}) = (\vec{x_1 - x_2, y_1 - y_2})$$

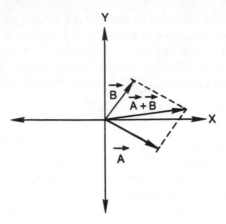

Fig. 1-17. Parallelogram method of adding vectors.

Sum and Difference of Vectors in Polar Coordinates

For polar coordinates, it is best to convert the vectors into Cartesian form, then add or subtract these values, and convert the resultant back into polar form.

Vectors can be multiplied in either of two ways: the *dot product* or the *cross product*. The dot product yields a real number, or scalar, while the cross product yields a vector that is perpendicular to the plane containing the original vectors. For vectors $\overrightarrow{A}$ and $\overrightarrow{B}$, let θ be the angle between the vectors, and A and B be their respective lengths (magnitudes).

Dot Product in Cartesian Coordinates

$$\overrightarrow{A} \bullet \overrightarrow{B} = x_1 x_2 + y_1 y_2$$

where $\overrightarrow{A} = \overrightarrow{(x_1, y_1)}$ and $\overrightarrow{B} = \overrightarrow{(x_2, y_2)}$.

Dot Product in Polar Coordinates

$$\overrightarrow{A} \bullet \overrightarrow{B} = AB \cos \theta$$

Cross Product in Cartesian Coordinates

$$\left| \overrightarrow{A} \times \overrightarrow{B} \right| = AB \sin \theta$$

where θ is determined according to

$$\theta = \arctan (y_2/x_2 - y_1/x_1),$$

and this gives the magnitude of the vector perpendicular to the two vectors **A** and **B**. The formula is the same in polar coordinates, and is in fact simpler since θ is determined simply by finding the difference between θ_B and θ_A, where these represent the angles that $\vec{B}$ and $\vec{A}$ subtend, respectively, between the axis $\theta = 0$.

Cross Product in Polar Coordinates

$$|A \times B| = AB \sin (\theta_B - \theta_A)$$

Notes about Cross Product

The cross product involves the introduction of a third dimension. If θ_B is greater than θ_A, the resultant vector will be as in the illustration. If θ_A is greater than θ_B, the resultant vector will be in the opposite direction.

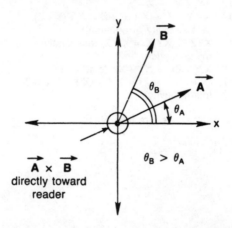

Fig. 1-18. Cross product of vector yields a perpendicular resultant.

The cross product is *not* commutative. That is, $\vec{A} \times \vec{B} \neq \vec{B} \times \vec{A}$. The dot product *is* commutative. That is, $\vec{A} \cdot \vec{B} = \vec{B} \cdot \vec{A}$.

The cross product is useful in determining the intensity and direction of propagation of an electromagnetic field.

1.23 BOOLEAN ALGEBRA

Boolean algebra is a system of mathematical logic, using the functions AND, NOT and OR. The AND function is represented by multiplication, NOT by complementation ('), and OR by addition.

Thus X and Y is written XY, NOT X is written X' (although sometimes you will see it denoted $-X$ or $\overline{X}$), and X OR Y is written $X + Y$. Some common rules for Boolean algebra are shown below.

1. $X + 0 = X$ (additive identity)
2. $X1 = X$ (multiplicative identity)
3. $X + 1 = 1$
4. $X0 = 0$
5. $X + X = X$
6. $XX = X$
7. $(X')' = X$ (double negation)
8. $X + X' = 1$
9. $X'X = 0$
10. $X + Y = Y + X$ (commutativity of addition)
11. $XY = YX$ (commutativity of multiplication)
12. $X + XY = X$
13. $XY' + Y = X + Y$
14. $X + Y + Z = (X + Y) + Z = X + (Y + Z)$
    (associativity of addition)
15. $XYZ = (XY)Z = X(YZ)$ (associativity of multiplication)
16. $X(Y + Z) = XY + XZ$ (distributivity)
17. $(X + W)(Y + Z) = XY + XZ + WY + WZ$
    (distributivity)

Fig. 1-19. Some Boolean algebra theorems.

Chapter 2

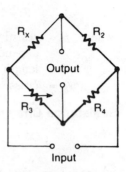

Resistance Formulas

The formulas given here for resistance and resistors cover the span from Ohm's law to expressions for multiple-resistor circuits. Resistance is universal in electronics, and the need to deal with it mathematically is ever present.

2.1 RESISTANCE AND RESISTIVITY

In electronics, *resistance* is a measure of the opposition to current flow in a circuit; it is a proportionality constant that describes the relationship between voltage and current. The term *resistivity* describes the bulk property of a material substance, such as carbon, which can be molded or formed into different shapes to make a component having a specific resistance value.

Ohm's Law

In an electrical circuit having a voltage E appearing across a resistance R, the current I flowing through the resistance is given by the formula

$$I = E/R$$

from which we can also determine that

$$E = IR$$

and

$$R = E/I$$

Resistance and Resistivity (Specific Resistance)

$$R = pl/A$$

where p = resistivity of conductor (ohm-cm or ohms/circ mil ft, depending upon units chosen for l and A)
l = length of conductor
A = cross-sectional area of conductor

Resistance Change Due to Temperature

$$R_{T2} = R_{T1}[1 + \alpha(T_2 - T_1)]$$

where T_1 = first temperature
T_2 = second temperature
R_{T1} = resistance at T_1
R_{T2} = resistance at T_2
α = temperature coefficient of material

Skin Effect

$$x = t\sqrt{(25.13\mu f)/(10^9 p)}$$

where t = thickness of conductor (cm)
f = frequency of current
μ = permeability of material
p = resistivity of material

Dynamic Resistance

$$r = \frac{de}{di}$$

That is, the dynamic resistance r is equal to the derivative of the voltage e with respect to the current i.

2.2 POWER DISSIPATED IN RESISTORS

All resistors consume and dissipate power as they resist the flow of current through them. The electrical power that is consumed by a resistor causes it to heat up, and as it does so, it dissipates the heat into the surrounding air or into anything else it comes into contact with (such as a heatsink). Since the ability of a resistor to dissipate heat is limited, the resistor is assigned a maximum power rating (in watts). The power actually dissipated by a resistor can be determined using any of the following three relationships:

where P = power, in watts
          R = resistance, in ohms
          E = voltage, in volts
          I = current, in amperes

Power When Both Voltage and Current Are Known

$$P = EI$$

Power When Both Current and Resistance Are Known

$$P = I^2 R$$

Power When Both Voltage and Resistance Are Known

$$P = E^2 / R$$

2.3 RESISTORS IN SERIES AND PARALLEL

Resistors may be combined in many different ways. The total resistance of the resulting combination can be calculated using the following formulas.

Resistors in Series

Total Resistance:

$$R_t = R_1 + R_2 + R_3 + \dots + R_n$$

Resistors in Parallel

For two resistors in parallel, the total resistance is

$$R_t = \frac{R_1 R_2}{R_1 + R_2}$$

For any number of resistors in parallel, the total resistance is

$$R_t = \cfrac{1}{\cfrac{1}{R_1} + \cfrac{1}{R_2} + \cfrac{1}{R_3} + \ldots + \cfrac{1}{R_n}}$$

Resistors in Series-Parallel

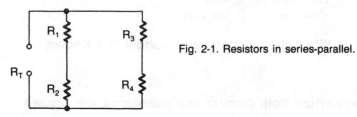

Fig. 2-1. Resistors in series-parallel.

Total resistance:

$$R_t = \cfrac{1}{\cfrac{1}{R_1 + R_2} + \cfrac{1}{R_3 + R_4}}$$

Resistors in Parallel-Series

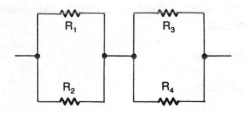

Fig. 2-2. Resistors in parallel-series.

Total resistance:

$$R_t = \cfrac{1}{\cfrac{1}{R_1} + \cfrac{1}{R_2}} + \cfrac{1}{\cfrac{1}{R_3} + \cfrac{1}{R_4}}$$

2.4 USEFUL RESISTOR CIRCUITS

There are many useful circuits that are constructed from resistors. Perhaps the most common is the voltage divider. In addition, there are circuits for dropping higher voltages down to lower voltages, and for changing the measurement ranges of ammeters and voltmeters.

Voltage Divider

$$E_{out} = E_{in}(R_2/R_1)$$

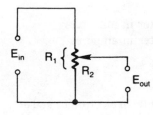

Fig. 2-3. Voltage divider.

Dropping Resistor

$$R_{drop} = (E_1 - E_2)/I$$

where E_1 = supply voltage
E_2 = desired load voltage
I = load current

Voltmeter Multiplier Resistor

$$R = E/I$$

where E = desired full-scale deflection voltage (volts)
I = full-scale deflection current of meter (amps)

Range-Increasing Resistor for Voltmeter

$$R = R_m(n - 1)$$

where R_m = total resistance of voltmeter
n = factor by which present full-scale deflection of meter is to be multiplied

Range-Increasing Resistor for Ammeter

$$R = R_m/(n - 1)$$

where R_m = internal resistance of meter, and
n = factor by which original current scale is to be multiplied

Internal Resistance of Moving-Coil Meter

$$R_m = 0.001 \, (E/I)$$

where E = full-scale deflection of meter in millivolts
I = full-scale deflection of meter in amperes

2.5 T AND PI NETWORKS

Both T and Pi networks occur frequently in electronics and are so called because of their resemblance to those letters. Basically, they are simple combinations of three resistors that have many useful circuit properties.

T Network

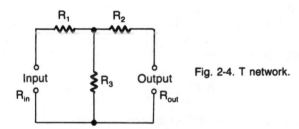

Fig. 2-4. T network.

Input resistance with output terminals unloaded:

$$R_{in} = R_1 + R_3$$

Output resistance with input terminals unloaded:

$$R_{out} = R_2 + R_3$$

Input resistance with output terminals short-circuited:

$$R_{in} = \frac{R_1 + R_2 R_3}{R_2 + R_3}$$

Output resistance with input terminals short-circuited:

$$R_{out} = \frac{R_2 + R_1 R_3}{R_1 + R_3}$$

Pi Network

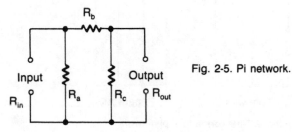

Fig. 2-5. Pi network.

Input resistance with output terminals unloaded:

$$R_{in} = \frac{R_a(R_b + R_c)}{R_a + R_b + R_c}$$

Output resistance with input terminals unloaded:

$$R_{out} = \frac{R_c(R_a R_b)}{R_a + R_b + R_c}$$

Input resistance with output terminals short-circuited:

$$R_{in} = \frac{R_a R_b}{(R_a + R_b)}$$

Output resistance with input terminals short-circuited:

$$R_{out} = \frac{R_b R_c}{R_b + R_c}$$

41

Transforming a T-Network into a Pi-Network

$$R_a = \frac{R_1R_2 + R_2R_3 + R_1R_3}{R_2}$$

$$R_b = \frac{R_1R_2 + R_2R_3 + R_1R_3}{R_3}$$

$$R_c = \frac{R_1R + R_2R_3 + R_1R_3}{R_1}$$

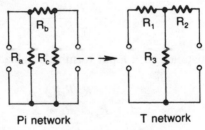

T network   Pi network

Fig. 2-6. Transforming a T-network into a pi-network.

Transforming a Pi-Network into a T-Network

Pi network   T network

Fig. 2-7. Transforming a pi-network into a T-network.

$$R_1 = \frac{R_aR_b}{R_a + R_b + R_c}$$

$$R_2 = \frac{R_bR_c}{R_a + R_b + R_c}$$

$$R_3 = \frac{R_aR_c}{R_a + R_b + R_c}$$

42

2.6 RESISTANCE BRIDGE

A resistance bridge circuit consists of four resistors connected as shown. The unique property of a bridge circuit is that the voltage appearing at the output goes to zero when the bridge is *balanced*. Resistance bridges may use either an ac or dc input voltage.

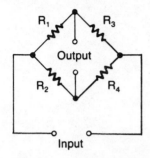

Fig. 2-8. Resistance bridge.

For a *null* (zero volts output), the resistances in the bridge circuit must be such that

$$R_1/R_2 = R_3/R_4$$

Any one of the resistances can be unknown, and its value determined in terms of the other three, as shown by the following formulas.

$$R_1 = R_2R_3/R_4$$
$$R_2 = R_1R_4/R_3$$
$$R_3 = R_1R_4/R_2$$
$$R_4 = R_2R_3/R_1$$

Chapter 3

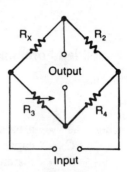

Capacitance
Formulas

The formulas in this chapter range from the basic property of capacitance to the capacitance of various combinations of capacitors. Capacitors are able to store electrical energy. Also, their reactance is not constant, but varies inversely with frequency. These unique properties of capacitance are very important in electronics.

3.1 CAPACITANCE OF PARALLEL PLATES

The simplest way to make a capacitor is to position two conductive plates so that their surfaces are parallel to each other, but do not touch each other. The capacitance value can be increased significantly by inserting dielectric insulating materials between the plates. In tubular capacitors, the metal foil and insulating strips are rolled up together, but for all practical purposes they may be considered to be a special type of parallel plate capacitor.

Capacitance of Two-Plate Capacitor

$$C = 0.2244 \ kA/d$$

where A = area of one plate (sq in.)
C = capacitance (pF)
d = distance between plates (in.)

45

$$k = \text{dielectric constant of dielectric between plates}$$
$$k = 1 \text{ for air}$$

When A is in square centimeters and d is in centimeters,

$$C = 0.08842 \ kA/d$$

Simplified Formulas for Two-Plate Capacitor

$$C = kA/(4.45 \ d)$$

where A = area of one plate (sq in.)
   C = capacitance (pF)
   D = distance between plates (in.)

For air dielectric, $k = 1$ and

$$C = A/(4.45d)$$

Capacitance of Multiple-Plate Capacitor

$$C = 0.2244 \ (n-1) \ kA/d$$

where n = number of plates
   A = area of one plate (sq in.)
   C = capacitance (pF)
   D = distance between plates (in.)
   k = dielectric constant of dielectric between plates

When A is in square centimeters and d is in centimeters,

$$C = 0.08842 \ (n-1) \ kA/d$$

Tubular Capacitors

When two strips of foil are alternately separated by two strips of dielectric insulation, and all the strips are rolled up tightly to form a tubular capacitor, the capacitance is given approximately by the formula:

$$C = 0.4488 \ kA/d$$

where A = average area of one foil strip (sq in.)
        C = capacitance (pF)
        d = thickness of dielectric insulation (in.)
        k = dielectric constant of insulation

3.2 CAPACITANCE OF STRAIGHT WIRES AND CABLES

Capacitance always exists between a single conductor and its surroundings, such as the earth, a metal chassis, or another conductor. Generally this capacitance is too small to be bothered with. However, at high frequencies, even small capacitances may have an appreciable effect on the operation of a circuit. The capacitance of coaxial cables is much larger owing to the close proximity of the shielding to the center conductor.

Capacitance of Single, Straight Wire Parallel to Ground

When height of the wire above ground is much greater than the diameter of the wire,

$$C = 7.354/\log_{10} (4h/D)$$

where C = capacitance in pF per foot
        D = diameter of wire (in.)
        h = height of wire above ground (in.)

Capacitance of Single, Straight Vertical Wire Above Ground

When the height of the bottom end of the wire above ground is much greater than the length l of the wire,

$$C = 0.2416 \, l \, (\log_{10} 2 \, l/D)$$

where C = capacitance (pF)
        D = diameter of wire (cm)
        l = length of wire (cm)

Capacitance Between Parallel Wires Above Ground

When the diameter of the wire is much smaller than the distance between wires,

$$C = 3.677/\log_{10}(2d/D)$$

where C = capacitance in pF per foot
d = distance between wires (in.)
D = diameter of wires (in.)

Capacitance of Concentric (Coaxial) Cable

$$C = 7.354 \ k \ \log_{10}(D_1/D_2)$$

where C = capacitance in pF per foot
D_1 = inside diameter of outer conductor (in.)
D_2 = outside diameter of inner conductor (in.)
k = dielectric constant of dielectric between conductors

3.3 CAPACITANCE OF DISCS AND SPHERES

The capacitance of disc and spherical shapes is often of interest. Here are some special formulas which apply to these special cases.

Capacitance of Two Parallel Discs

$$C = \frac{kD^2}{5.666d}$$

where C = capacitance (pF)
D = diameter of disc (in.)
d = distance between discs (in.)
k = dielectric constant of dielectric between discs

Capacitance of Single, Thin Isolated Disc

When a disc is placed at the center of a chamber whose dimensions are much larger than the diameter of the disc, the disc exhibits a capacitance:

$$C = 0.8992 \ kD$$

where C = capacitance (pF)

D = diameter of disc (in.)
k = dielectric constant of dielectric filling chamber
k = 1 for air

Capacitance of Two Concentric Spheres

$$C = 1.412\ k\ D_1 D_2/(D_2 - D_1)$$

where C = capacitance (pF)
D_1 = outer diameter of inside sphere (in.)
D_2 = inside diameter of outside sphere (in.)
k = dielectric constant of dielectric between spheres

Capacitance of Single, Isolated Sphere

When a sphere is placed at the center of a chamber whose dimensions are much larger than diameter of the sphere, the sphere exhibits a capacitance:

$$C = 1.412\ kD$$

where C = capacitance (pF)
D = diameter of sphere (in.)
k = dielectric constant of dielectric in chamber
k = 1 for air

3.4 CAPACITOR CHARGE AND ENERGY

Capacitors, like inductors, are energy-storage devices. This means that capacitors are able to store energy as an electric charge is built up between the plates of the capacitor. The fundamental relationships are given below.

Charge Stored in a Capacitor

$$Q = C\ E$$

where Q = electric charge (coulombs)
C = capacitance (F)
E = voltage (V)

Variations of the above formula permit us to determine the voltage across a charged capacitor, $E = Q/C$, and the capacitance when the charge is known, $C = Q/E$.

Energy Stored in A Capacitor

$$W = \tfrac{1}{2}CE^2$$

where W = energy in watt-seconds (joules)
         C = capacitance (F)
         E = capacitor voltage (V)

3.5 CAPACITOR CURRENT, VOLTAGE, AND REACTANCE

The relationship between current and voltage in a capacitor are somewhat more complex than for a simple resistor. Also, the manner in which a capacitor (or inductor) resists the flow of current is referred to as *reactance*, rather than resistance. However, both reactance and resistance are measured in ohms.

Direct Current In or Out of a Capacitor

$$I = \frac{dQ}{dt} = C\,\frac{dE}{dt}$$

where I = current (A)
   dQ/dt = derivative of the charge stored in the capacitor with respect to time t
       C = capacitance (F)
   dE/dt = derivative of the capacitor voltage E with respect to time t

If the capacitor voltage is changing at a constant rate, then the current effectively flowing through the capacitor may be determined by the following formula.

$$I = C(E_2 - E_1)/(t_2 - t_1)$$

where E_2 = capacitor voltage at time t_2
      E_1 = capacitor voltage at time t_1

Direct Voltage Across a Capacitor

$$E = \frac{1}{C} \int I \, dt$$

where E = capacitor voltage (V)
C = capacitance (F)
I = capacitor current (A)

If the current flowing into or out of the capacitor is constant, then the integral simplifies and

$$E_2 - E_1 = I(t_2 - t_1)/C$$

where E_2 = capacitor voltage at time t_2
E_1 = capacitor voltage at time t_1

Capacitive Reactance

$$X_C = 1/(2\pi f C)$$

where X_C = capacitive reactance (ohms)
f = frequency of sine-wave voltage or current (Hz)
C = capacitance (F)
π = 3.14159

Capacitance in Terms of Reactance

$$C = 1/(2\pi f X_C)$$

where X_C = capacitive reactance (ohms)
f = frequency (Hz)

Capacitance in Terms of Resonant LC Circuit

$$C = 1/(4\pi^2 f^2 L)$$

where f = resonant frequency (Hz)
L = resonating inductance (H)

Alternating Current Through a Capacitor

$$I = E/X_C$$

where I = capacitor current (A)
E = capacitor voltage (V)
X_C = capacitive reactance (ohms)

Capacitor Figure of Merit (Q)

$$Q = X_C/R$$

where Q = figure of merit (the higher the better)
X_C = capacitive reactance (ohms)
R = effective series resistance (ESR) of capacitor (ohms)

Alternating Voltage Across a Capacitor

$$E = I X_C$$

where I = capacitor current (A)
E = capacitor voltage (V)
X_C = capacitive reactance (ohms)

Phase Angle Between Current and Voltage in a Capacitor

$$\theta_1 - \theta_E = 90 \text{ degrees}$$

where θ_1 = phase angle of sine-wave current
θ_E = phase angle of sine-wave voltage

3.6 CAPACITORS IN SERIES AND PARALLEL

Capacitors may be combined in a variety of ways. Capacitive reactances always combine in the same way that resistances do. However, capacitance values must be handled somewhat differently than resistances:

Capacitors in Parallel

Total capacitance:

$$C_t = C_1 + C_2 + C_3 + \ldots + C_n$$

52

Capacitors in Series

Total capacitance of any number in series:

$$C_t = \frac{1}{\dfrac{1}{C_1} + \dfrac{1}{C_2} + \dfrac{1}{C_3} + \ldots + \dfrac{1}{C_n}}$$

Total capacitance of only two in series:

$$C_t = (C_1 C_2)/(C_1 + C_2)$$

Capacitors in Series—Parallel

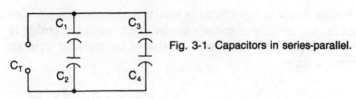

Fig. 3-1. Capacitors in series-parallel.

Total capacitance:

$$C_t = \frac{1}{\dfrac{1}{C_1} + \dfrac{1}{C_2}} + \frac{1}{\dfrac{1}{C_3} + \dfrac{1}{C_4}}$$

Capacitors in Parallel—Series

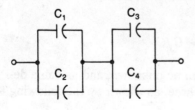

Fig. 3-2. Capacitors in parallel-series.

Total capacitance:

$$C_t = \frac{1}{\dfrac{1}{C_1 + C_2} + \dfrac{1}{C_3 + C_4}}$$

53

Capacitive Voltage Divider

$$E_{out} = \frac{E_{in}C_1}{C_1 + C_2}$$

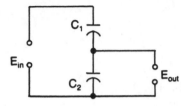

Fig. 3-3. Capacitive voltage divider.

3.7 CAPACITANCE BRIDGE

A capacitance bridge circuit is similar to the resistance bridge circuit in that the output voltage goes to zero when the bridge is balanced. Capacitance bridges must always be operated from an ac signal source.

Fig. 3-4. Capacitance bridge.

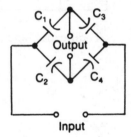

For null (zero volts output), the capacitances in the bridge circuit must be such that

$$C_1/C_2 = C_3/C_4$$

Any one of the capacitors can be unknown, and its value determined in terms of the other three, as shown by the following formulas.

$$C_1 = (C_2C_3)/C_4$$
$$C_2 = (C_1C_4)C_3$$
$$C_3 = (C_1C_4)/C_2$$
$$C_4 = (C_2C_3)/C_1$$

Chapter 4

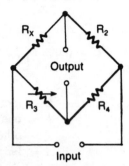

Inductance Formulas

The formulas included in this chapter range from the determination of coil and straight-wire inductances, through basic inductance equations. Inductors are able to store electrical energy in the magnetic field they produce. Also, their reactance varies proportionally with frequency. The properties of inductance and capacitance complement each other and thus permit the construction of very complex and useful electronic circuits.

4.1 INDUCTANCE OF STRAIGHT WIRES AND CABLES

The flow of electric current through a wire creates a magnetic field about the wire. The inductance of a wire or cable is a measure of how much of a magnetic field is produced for a given amount of current. The inductance of wires and cables can be an important consideration at radio frequencies.

Inductance of Straight Round Wire

When length l is at least 1000 times the diameter d of the wire,

$$L = 0.002\, l\, [2.303 \log_{10} (4\, l/d) - 0.75]$$

where L = inductance (μH)
        l = length (cm)
        d = diameter (cm)

When $l < 1000d$,

$$L = 0.002\ l\ [2.303\ \log_{10}\ (4\ l/d) - 0.75(+d/2l)]$$

Inductance of Straight Rectangular Conductor

$$L = 0.002\ l\left[2.303\ \log_{10}\left(\frac{2\ l}{b + c}\right) + 0.5 + 0.2235(b + c)/l\right]$$

where L = inductance (μH)
$\quad\quad\quad b$ = one side of rectangle (cm)
$\quad\quad\quad c$ = other side of rectangle (cm)
$\quad\quad\quad l$ = length of conductor (cm)

Inductance of Two Parallel,
Round Wires (One Forming Return Circuit)

$$L = 0.004\ l\ [2.303\ \log_{10}(2\ D/d - D/l)]$$

where L = inductance (μH)
$\quad\quad\quad D$ = distance between centers of wires (cm)
$\quad\quad\quad d$ = diameter of wire (cm)
$\quad\quad\quad l$ = length of conductor (cm)

Inductance of Concentric (Coaxial) Cable

$$L = 0.140\ \log_{10}(r_2/r_1) + 0.015$$

where L = inductance (μH/ft)
$\quad\quad\quad r_1$ = outside radius of inner conductor (in.)
$\quad\quad\quad r_2$ = inside radius of outer conductor (in.)

4.2 INDUCTANCE OF SOLENOIDS AND COILS

The most efficient way to obtain high inductance values is to wind the wire into a coil or solenoid. The inductance may be further increased by winding the coil about a magnetic core.

Inductance of Single-Layer Solenoid (Without Magnetic Core)

$$L = \frac{0.2\, d^2 N^2}{3d + 9l}$$

where L = inductance (μH)
      d = diameter of winding (in.)
      l = length of winding (in.)
      N = number of turns

Inductance of Multilayer Solenoid (Without Magnetic Core)

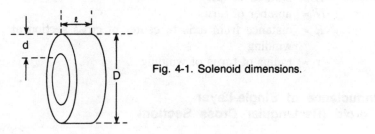

Fig. 4-1. Solenoid dimensions.

$$L = \frac{0.2\, D^2 N^2}{3D + 9l + 10d}$$

where L = inductance (μH)
      D = diameter of coil (in.)
      l = length of winding (in.)
      d = radial depth of coil (in.)
      N = number of turns

Inductance of Coil With Magnetic Core

$$L = \frac{4.06\, N^2 \mu A}{(1.27 \times 10^8)\, l}$$

where L = inductance (H)
      l = total length of core (in.)
      A = cross-sectional area of core (sq in.)
      N = number of turns
      μ = permeability of core material

Inductance of Single-Layer Toroid (Circular Cross Section)

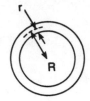

Fig. 4-2. Cross section of a toroid.

$$L = 0.01257 \, N^2 \, (R - \sqrt{R^2 - r^2})$$

where L = inductance (μH)
  N = number of turns
  R = distance from axis to center of cross section of winding
  r = radius of turns of winding

Inductance of Single-Layer Toroid (Rectangular Cross Section)

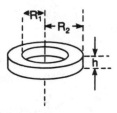

Fig. 4-3. Toroid cross section (dimensions).

$$L = 0.004606 \, N^2 h \, \log_{10}(R_2/R_1)$$

where L = inductance (μH)
  R_1 = inner radius (cm)
  R_2 = outer radius (cm)
  N = number of turns
  h = axial depth (cm)

Inductance of Multilayer Square Coil (With Square Cross Section)

$$L = 0.008 \, A \, N^2 \, [2.303 \, \log_{10}(A/B) + 0.477(B/A) + 0.033]$$

where L = inductance (μH)
A = side of square measured to center of square cross section (cm)
B = one side of square winding (cm)
N = number of turns

4.3 MUTUAL INDUCTANCE AND COUPLING

When two wires or coils are brought close to each other, their magnetic fields overlap so that the inductance of one affects the inductance of the other. This effect is known as *mutual* inductance.

Mutual Inductance Between Two Coils

When two coils have identical cores or are wound on the same core,

$$M = \frac{4.06 \ N_1 N_2 \ \mu \ A}{1.27 \times 10^8 \ l}$$

where M = mutual inductance (H)
N_1 = number of turns in first coil
N_2 = number of turns in second coil
A = cross-sectional area of core (sq in.)
l = total length of core (in.)
μ = permeability of core material

Mutual Inductance Between Two Parallel Wires of Equal Length

When the distance D between wires is much greater than the diameter d of wire,

$$M = 0.002 \ l \ [(2.303 \ \log_{10}(2 \ l/D) - (1 + D) \ /l \ ]$$

where M = mutual inductance (H)
D = distance between wires (cm)
d = diameter of wire (cm)
l = length of wire (cm)

Coefficient of Coupling Between Two Coils

$$k = M/\sqrt{L_1 L_2}$$

where L_1 = inductance of first coil
L_2 = inductance of second coil
M = mutual inductance between coils

4.4 INDUCTOR FIELD AND ENERGY

An inductor stores energy in its magnetic field, much as a capacitor is able to store energy in its electric field. The fundamental relationships are given below.

Magnetic Field in an Inductor

$$\Phi = LI$$

where Φ = magnetic flux (webers)
L = inductance (H)
I = current (A)

Energy Stored in Magnetic Field of Inductor

$$W = \tfrac{1}{2} L I^2$$

where W = energy (watt-sec)
L = inductance (H)
I = current flowing through inductor (A)

4.5 INDUCTOR CURRENT, VOLTAGE, AND REACTANCE

As with the capacitor, the inductor has a somewhat complex relationship between current, voltage, and reactance. The *reactance* of an inductor is a measure of the amount which the inductor resists the flow of current. Reactance is measured in units of ohms.

$$E = \frac{d\Phi}{dt} = L \frac{dI}{dt}$$

where E = voltage applied across inductor
$d\Phi/dt$ = derivative of magnetic field Φ stored in inductor with respect to time t
L = inductance (H)
dI/dt = derivative of inductor current with respect to time

If the inductor current is changing at a constant rate, then the voltage applied across the inductor may be determined from the following formula:

$$E = L(I_2 - I_1)/(t_2 - t_1)$$

where I_2 = inductor current at time t_2
I_1 = inductor current at time t_1

Counter EMF (Back Voltage) of Inductor

An inductor resists any attempt to change the amount of current flowing through it by generating an opposing voltage.

$$E = -L\frac{dI}{dt}$$

where E = voltage generated by inductor
dI/dt = derivative of inductor current I with respect to time t
L = inductance (H)

Alternating Current Through Inductor

$$I = \frac{1}{L}\int E\,dt$$

where I = inductor current (A)
L = inductance (H)
E = inductor voltage (V)

If the voltage across the inductor is constant, then the integral simplifies so that

$$I_2 - I_1 = E(t_2 - t_1)/L$$

where I_2 = inductor current at time t_2
I_1 = inductor current at time t_1

Inductive Reactance

$$X_L = 2\pi f L$$

where X_L = inductive reactance (ohms)
f = frequency of sine-wave voltage or current (Hz)
L = inductance (H)
π = 3.14159

Inductance in Terms of Reactance

$$L = 2\pi f X_L$$

where f = frequency
X_L = inductive reactance (ohms)

Inductance in Terms of Resonant LC Circuit

$$L = 1/(4\pi^2 f^2 C)$$

where f = frequency (Hz)
C = capacitance (F)

Inductor Figure of Merit (Q)

$$Q = X_L/R$$

where Q = figure of merit (the higher the better)
X_L = inductive reactance (ohms)
R = series resistance of inductor (ohms)

Alternating Current Through Inductor

$$I = E/X_L$$

where I = current (A)
E = voltage (V)
X_L = inductive reactance (ohms)

Alternating Voltage Across Inductor

$$E = IX_L$$

62

where E = voltage (V)
I = current (A)
X_L = inductive reactance (ohms)

Phase Angle Between Current and Voltage in an Inductor

$$\theta_E - \theta_1 = 90 \text{ degrees}$$

where θ_E = phase angle of sine-wave voltage
θ_1 = phase angle of sine-wave current

4.6 INDUCTORS IN SERIES AND PARALLEL

Inductors may be combined in a variety of ways. Inductive reactances always combine in the same way that resistances do. To a certain extent, inductance values combine like resistance values. However, combining inductances is often complicated by the effects of the mutual inductance between inductors.

Inductors in Series (Simple Case)

Neglecting mutual inductance, the total inductance is

$$L_t = L_1 + L_2 + L_3 + \dots + L_n$$

Inductors in Series (Exhibiting Mutual Inductance)

Total Inductance:

$$L_t = L_1 + L_2 + 2M$$

where L_1 = self-inductance of first coil
L_2 = self-inductance of second coil
M = mutual inductance between coils

If the connections to one coil are reversed, the total inductance is

$$L_t = L_1 + L_2 - 2M$$

Inductors in Parallel

Neglecting mutual inductance, the total inductance is

$$L_t = \cfrac{1}{\cfrac{1}{L_1} + \cfrac{1}{L_2} + \cfrac{1}{L_3} + ... + \cfrac{1}{L_n}}$$

Inductors in Series-Parallel

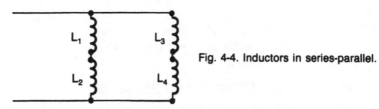

Fig. 4-4. Inductors in series-parallel.

Neglecting mutual inductance, the total inductance is

$$L_t = \cfrac{1}{\cfrac{1}{L_1 + L_2} + \cfrac{1}{L_3 + L_4}}$$

Inductors in Parallel-Series

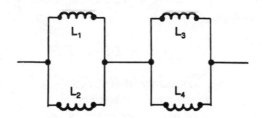

Fig. 4-5 Inductors in parallel-series.

Neglecting mutual inductance, the total inductance is

$$L_t = \frac{1}{L_1 + L_2} + \frac{1}{L_3 + L_4}$$

Inductive Voltage Divider

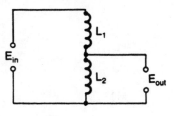

Fig. 4-6. Inductive voltage divider.

$$E_{out} = E_{in}L_2/(L_1 + L_2)$$

4.7 INDUCTANCE BRIDGE

An inductance bridge is similar to a capacitance bridge in that it must be operated from an ac signal source. The output voltage of the bridge circuit goes to zero when the bridge is balanced.

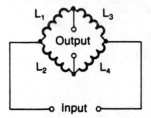

Fig. 4-7. Inductance bridge.

For null (zero volts output), the inductance values must be such that

$$L_1/L_2 = L_3/L_4$$

Any one of the inductances can be unknown, and its value determined in terms of the other three, as shown by the following formulas.

$$L_1 = L_2L_3/L_4$$
$$L_2 = L_1L_4/L_3$$
$$L_3 = L_1L_4/L_2$$
$$L_4 = L_2L_3/L_1$$

Chapter 5

Formulas for Charge, Current, Voltage, and Power

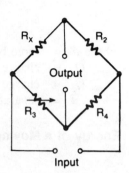

This chapter covers the relationships between charge, current, voltage, and power in electric circuits. These formulas form the basis for the electric circuit formulas given in the next chapter.

5.1 CHARGE AND ELECTRONS

Electric current is composed of moving electrons, each of which possesses a small electric charge. The basic unit of electric charge Q is the *coulomb*, which is approximately equal to 6.242×10^{18} electrons. This means that each electron possesses an electric charge of 1.602×10^{-19} coulombs.

Charge in Terms of Current and Time

$$Q = \int I \, dt = It$$

where Q = electric charge (coulombs)
$\int I \, dt$ = integral of the current I in amperes with respect to time t in seconds

If the current flow is constant, then the integral simplifies so that

$$Q_2 - Q_1 = I(t_2 - t_1)$$

where Q_2 = charge at time t_2
Q_1 = charge at time t_1

Coulomb's Law

$$F = (Q_1 Q_2)/d^2$$

where F = force between two charged bodies in air
Q_1 = charge on first body
Q_2 = charge on second body
d = separation between bodies (cm)

Energy of a Moving Electron

$$U = \frac{1}{2}mv^2$$

where U = energy (ergs)
m = mass of an electron (9.109×10^{-28} gram)
v = velocity of electron (cm/sec)

5.2 DIRECT CURRENT

The basic unit for current is the *ampere,* which is defined as the flow of one coulomb of electric charge per second. The term *direct current* (dc) commonly refers to current which flows in one direction only. (Sometimes the term dc is used to refer to a static or unchanging voltage, while the term ac refers to the dynamic or changing component.)

Current in Terms of Quantity and Time

$$I = \frac{dQ}{dt}$$

where dQ/dt = derivative of charge Q with respect to time t

If the charge rate is constant, then the current flow can be obtained from the following formula:

$$I = (Q_2 - Q_1)/t_2 - t_1)$$

where Q_2 = charge at time t_2
Q_1 = charge at time t_1

Current Charge or Discharge

$$Ah = It$$

where Ah = total charge or drain (ampere-hours)
I = current (amperes)
t = total time (hours)

Current in Terms of Voltage and Resistance

$$I = E/R$$

Current in Terms of Voltage and Power

$$I = P/E$$

Current in Terms of Power and Resistance

$$I = \sqrt{P/R}$$

5.3 DIRECT VOLTAGE

It is more common to refer to direct voltage as dc voltage, signifying a voltage that arises as a result of the flow of a direct current (dc). In general usage the term dc voltage refers to a voltage having a single polarity, being either positive or negative with respect to some reference point.

Voltage in Terms of Current and Resistance

$$E = IR$$

Voltage in Terms of Current and Power

$$E = P/I$$

Voltage in Terms of Power and Resistance

$$E = \sqrt{PR}$$

5.4 ALTERNATING CURRENT

Alternating current (ac) differs from direct current (dc) only in that it alternately changes direction of polarity. Unless stated otherwise, the term ac normally implies a sinusoidal waveform.

Instantaneous Alternating Current

$$i = I_m \sin \theta$$

where I_m = maximum or peak value of current
θ = phase angle

When the frequency f or the angular velocity ω is given, the following formula may be used.

$$i = I_m \sin \omega t$$

where $\omega = 2\pi f$
t = time (sec)

Average Alternating Current

$$I_{avg} = 0.637 \, I_{max}$$

where I_{max} = maximum current

Effective (RMS) Alternating Current

$$I_{eff} = I_{rms} = 0.707 \, I_{max}$$

where I_{max} = maximum current

Current in Terms of Voltage and Reactance

$$I = E/X_C = E/(2\pi f C)$$

$$I = E/X_L = E/(2\pi fL)$$

where I = current (A)
E = voltage (V)
X_C = capacitive reactance (ohms)
X_L = inductive reactance (ohms)
f = frequency of sine-wave voltage (Hz)
C = capacitance (F)
L = inductance (H)

Current in Terms of Voltage and Impedance

$$I = E/Z = E/\sqrt{R^2 + X^2}$$

where I = current (A)
E = voltage (V)
Z = impedance (ohms)
R = resistance (ohms)
X = reactance (ohms)

5.5 ALTERNATING VOLTAGE

It is more common to speak of ac voltage rather than alternating voltage, denoting that the voltage arises from an alternating current (ac). Unless stated otherwise, ac voltage normally implies that the voltage has a sinusoidal waveform.

Instantaneous Ac Voltage

$$e = E_m \sin \theta$$

where E_m = maximum voltage
θ = phase angle

When the frequency f or the angular velocity ω is given, the following formula may be used:

$$e = E_m \sin \omega t$$

where ω = $2\pi f$
t = time (sec)

Average Ac Voltage

$$E_{avg} = 0.637 \, E_m$$

where E_m = maximum voltage

Effective (RMS) Ac Voltage

$$E_{eff} = E_{rms} = 0.707 \, E_m$$

where E_m = maximum voltage

Voltage in Terms of Current and Reactance

$$E = IX_C = I/(2\pi fC)$$
$$E = IX_L = 2\pi fLI$$

where I = current (A)
E = voltage (V)
X_C = capacitive reactance (ohms)
X_L = inductive reactance (ohms)
f = frequency of sine-wave current (Hz)
C = capacitance (F)
L = inductance (H)

Voltage in Terms of Current and Impedance

$$E = IZ = I\sqrt{R^2 + X^2}$$

where E = voltage (V)
I = current (A)
Z = impedance (ohms)
R = resistance (ohms)
X = reactance (ohms)

Voltage Reduction Produced By Series Resistor

$$E_{out} = E_{in} - I_L R$$

where E_{out} = output voltage across load

$$E_{in} = \text{input supply voltage}$$
$$I_L = \text{load current}$$
$$R = \text{series dropping resistance}$$

Voltage Regulation

$$VR = 100(E_1 - E_2)/E$$

where VR = voltage regulation (%)

E_1 = no-load voltage

E_2 = full-load voltage

5.6 POWER

Electric power is expressed in several different units. Real or effective power is expressed in *watts* (W). Imaginary or reactive power is expressed in *volt-amperes reactive* (var) units. Apparent power is expressed in (VA), which is simply the product of the line voltage and line current. Real and imaginary power are also determined from the product of voltage and current, but the phase angle between the two waveforms must be taken into account. The cosine of the phase angle is referred to as the power factor (pf).

Dc Power

$$P = IE$$
$$P = I^2R$$
$$P = E^2/R$$

where P = real or effective power (W)

I = current (A)

E = voltage (V)

R = resistance (ohms)

Ac Power

$$P = EI \cos \theta$$
$$P = I^2R \cos \theta$$
$$P = (E^2/R) \cos \theta$$

73

where P = real or effective power (W)
  E = voltage (V)
  I = current (A)
  R = resistance (ohms)
  θ = phase angle between current and voltage waveforms
  $\cos \theta$ = 1 when θ = 0

Reactive (Wattless) Power

$$P = I^2 X_C$$
$$P = I^2 X_L$$
$$P = E^2 / X_C$$
$$P = E^2 / X_L$$
$$P = EI \sin \theta$$

where P = imaginary or reactive power (var)
  I = current (A)
  E = voltage (V)
  X_L = inductive reactance (ohms)
  X_C = capacitive reactance (ohms)
  θ = phase angle between voltage and current waveforms
  $\sin \theta$ = 1 when θ = 90 degrees

Apparent Power (volt-amps)

$$P_a = EI = \sqrt{P_r^2 + P_i^2}$$

where P_a = apparent power (VA)
  P_r = real or effective power (W)
  P_i = imaginary or reactive power (var)
  E = voltage (V)
  I = current (A)

Power Factor (pf)

$$\text{pf} = \cos \theta$$
$$\text{pf} = R/Z = R/\sqrt{R^2 + X^2}$$

74

where θ = phase angle between voltage and current waveforms

$\cos \theta$ = 1 when $\theta = 0$

$\cos \theta$ = 0 when $\theta = 90$ degrees

R = resistance (ohms)

Z = impedance (ohms)

X = reactance (ohms)

Power Efficiency

$$\eta_p = 100\, P_o/P_i$$

where η_p = power efficiency (%)

P_i = input power

P_o = output power

Energy Consumption

$$W = Pt$$
$$W = EIt$$
$$W = E^2t/R$$
$$W = I^2Rt$$

where W = energy in watt-hours (Wh)

P = power (W)

t = total time in hours (h)

E = voltage (V)

I = current (A)

When the current and voltage are unsteady, an integration is required:

$$W = \int EI\, dt$$

5.7 DECIBELS

It is often convenient to express the ratio of two voltages, two currents, or two powers in *decibels* (dB). In amplifiers such ratios are used to measure gain and attenuation. When the ratio is less than one, the decibel number will be negative.

Voltage Ratio in Decibels

$$N_{dB} = 20 \log_{10}(E_{out}/E_{in})$$

where E_{out} = output voltage
E_{in} = input voltage

Current Ratio in Decibels

$$N_{dB} = 20 \log_{10}(I_{out}/I_{in})$$

where I_{out} = output current
I_{in} = input current

Power Ratio in Decibels

$$N_{dB} = 10 \log_{10}(P_{out}/P_{in})$$

where P_{out} = output power
P_{in} = input power

Measuring Voltage, Current, and Power in Decibels

It is also possible to measure voltage, current, power, and other parameters in decibels. In this case, no ratio of terms is actually involved. However, it usually turns out that some reference value is always implied; for example, when measuring a voltage in decibels, the implied reference value is 1V, since $\log(V/1) = \log V$.

$$V_{dB} = 20 \log_{10} V$$
$$I_{dB} = 20 \log_{10} I$$
$$P_{dB} = 10 \log_{10} P$$

where V_{dB} = voltage in decibels, referred to 1 volt
I_{dB} = current in decibels, referred to 1 ampere
P_{dB} = power in decibels, referred to 1 watt

Power is expressed in milliwatts (mW) often enough that the

abbreviation dBm is used to mean that the power is given in decibels with respect to 1 mW. Thus, 3 dBm refers to a power level that is 3 dB above 1 milliwatt.

5.8 POLYPHASE POWER SYSTEMS

Two-phase and three-phase power distribution systems are quite common. The following formulas summarize the significant characteristics of these systems.

Three-Phase Line Voltage, Balanced Y

$$E_L = \sqrt{3}E_P \cong 1.732\ E_P$$

where E_P = voltage of one phase

Three-Phase Line Current, Balanced Delta

$$I_L = \sqrt{3}I_p \cong 1.732\ I_p$$

where I_p = current in one phase

Total Power in Three-Phase System

$$P_t = P_1 + P_2 + P_3$$

where P_1, P_2, and P_3 are power in individual phases.

If the three phases are balanced, such that $P_1 = P_2 = P_3$, then

$$
\begin{aligned}
P_t &= 3P_p \\
&= 3E_pI_p \cos \theta \\
&= \sqrt{3}E_LI_L \cos \theta
\end{aligned}
$$

where P_p = power in individual phase
E_p = voltage in one phase
I_p = current in one phase
θ = phase angle

Power in Individual Phase of Three-Phase System

$$P_p = E_pI_p \cos \theta$$

Two-Phase Line Voltage

$$E_L = \sqrt{2}E_p$$

where E_p = voltage in individual phase

Neutral Current in Two-Phase System

$$I_n = \sqrt{2}I_p$$

where I_p = current in individual phase

Note that the neutral current in a two-phase system is zero when the two line currents are balanced; that is, the two currents are equal and 180 degrees out of phase with each other.

Chapter 6

Basic
Circuit Formulas

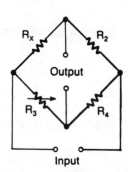

This chapter deals primarily with common circuits that contain combinations of resistors, capacitors, and inductors. Such circuits are described in terms of their impedance, phase angle, time constant, and frequency response.

6.1 IMPEDANCE

Impedance is a combination of resistance and reactance. The formulas involving impedance must take into account the fact that resistance and reactance are 90 degrees out of phase with each other. Consequently, resistance and reactance can only be combined using the rules of vector addition.

Since capacitive and inductive reactance are of opposite phase, they also have opposite signs.

Impedance of Series Resistance and Capacitance

$$Z = \sqrt{R^2 + X_C^2}$$

Impedance of Series Resistance and Inductance

$$Z = \sqrt{R^2 + X_L^2}$$

Impedance of Series Inductance and Capacitance

$$Z = X_L - X_C$$

At resonance, $X_L = X_C$, so that $Z = 0$.

Impedance of Parallel Resistance and Capacitance

$$Z = \frac{RX_C}{\sqrt{R^2 + X_C^2}}$$

Impedance of Parallel Resistance and Inductance

$$Z = \frac{RX_L}{\sqrt{R^2 + X_L^2}}$$

Impedance of Parallel Inductance and Capacitance

$$Z = \frac{X_L}{1 - \omega LC}$$

At resonance, $Z = \infty$.

Impedance of Series Resistance, Capacitance, and Inductance

$$Z = \sqrt{R^2 + (X_L - X_C)^2}$$

Impedance of Parallel Resistance, Capacitance, and Inductance

$$Z = \frac{RX_L X_C}{\sqrt{R(X_L - X_C)^2 + (X_L^2 X_C^2)}}$$

6.2 PHASE ANGLES

In all but the simplest circuits, the current and voltage waveforms are not precisely in phase with each other. This difference is expressed by stating the phase angle. The following formulas permit direct calculation of the phase angle using the known resistance and reactance values of the circuit components. The phase angle is normally assigned a positive value when the voltage waveform *leads* the current waveform and a negative value when the voltage *lags* the current.

Phase Angle of Series Resistance and Capacitance

$$\theta = -\tan^{-1}(X_C/R)$$

Phase Angle of Series Resistance and Inductance

$$\theta = \tan^{-1}(X_L/R)$$

Phase Angle of Series Inductance and Capacitance (R = 0)

$$\theta = 0 \text{ at resonance when } X_L = X_C$$
$$\theta = +90° \text{ when } X_L > X_C$$
$$\theta = -90° \text{ when } X_L < X_C$$

Phase Angle of Parallel Resistance and Capacitance

$$\theta = -\tan^{-1}(R/X_C)$$

Phase Angle of Parallel Resistance and Inductance

$$\theta = \tan^{-1}(R/X_L)$$

Phase Angle of Parallel Inductance and Capacitance

$$\theta = 0 \text{ at resonance when } X_L = X_C$$
$$\theta = +90° \text{ when } X_L < X_C$$
$$\theta = -90° \text{ when } X_L > X_C$$

Phase Angle of Series Resistance, Capacitance, and Inductance

$$\theta = \tan^{-1}\left(\frac{X_L - X_C}{R}\right)$$

$$\theta = 0 \text{ at resonance}$$

Phase Angle of Parallel
Resistance, Capacitance, and Inductance

$$\theta = \tan^{-1} \frac{RX_L - RX_C}{X_L X_C}$$

$\theta = 0$ at resonance

6.3 TIME CONSTANTS

The time constant of a circuit may be quickly determined for simple circuits using the following two formulas.

Time Constant of Capacitive-Resistive Circuit

$$\tau = RC$$

where τ = time constant (sec)
        R = resistance (ohms)
        C = capacitance (F)

Time Constant of Inductive-Resistive Circuit

$$\tau = L/R$$

where τ = time constant (sec)
        L = inductance (H)
        R = series resistance (including resistance of inductor)

6.4 CIRCUIT RESPONSE TO STEP INPUT

The response of a simple circuit to a step voltage input is summarized by the following formulas.

Current Flow in a Capacitive-Resistive Circuit

$$I = (E/R)\epsilon^{-t/RC}$$

where I = current (A)
        E = step voltage input (V)
        R = resistance in series with capacitor (ohms)
        C = capacitance (F)
        t = time elapsed since applying E (sec)

Current Flow in an Inductive-Resistive Circuit

$$I = (E/R)\epsilon^{-Rt/L}$$

where I = current (A)
E = step voltage input (V)
R = series resistance, including resistance of inductor (ohms)
L = inductance (H)
t = time elapsed since applying E (sec)

6.5 RESONANT FREQUENCY OF INDUCTANCE-CAPACITANCE CIRCUIT

$$f_r = \frac{1}{2\pi\sqrt{LC}}$$

From which:

$$C = 1/(4\pi^2 f^2 L)$$
$$L = 1/(4\pi^2 f^2 C)$$

If the resistive components of the capacitor and inductor are significant, then

$$f_r = \frac{L - R_L^2 C}{2\pi\sqrt{LC(L - R_C^2 C)}}$$

where R_L = series resistance of inductor
R_C = effective series resistance of capacitor

If the resistance of the inductor is the only significant resistive component, then

$$f_r = \frac{1 - R_L^2 C/L}{2\pi\sqrt{LC}}$$

Bandwidth of Tuned Circuit

$$\underline{B} = f_r/Q = f_2 - f_1$$

where B = bandwidth (Hz)
f_r = resonant frequency (Hz)
Q = figure of merit, X/R
f_1 = lower 3 dB cutoff frequency (Hz)
f_2 = upper 3 dB cutoff frequency (Hz)

6.6 FILTER DESIGN FORMULAS

The two most common filter types are the *constant-k* and the *m-derived*. These filters do not contain resistors, but they are intended to be terminated with resistances at their input and output. The formulas provided in this section cover the design of basic low-pass, high-pass, bandpass, and bandstop filters.

Constant-k Low-Pass Filter

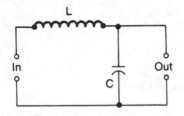

Fig. 6-1. Constant-k low-pass filter.

$$L = R/\pi f$$
$$C = 1/\pi f R$$

where f = cutoff frequency
R = terminating resistance

Series M-Derived Low-Pass Filter

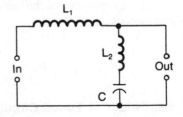

Fig. 6-2. Series m-derived low-pass filter.

$$m = \sqrt{1 - (f_1^2/f_2^2)}$$
$$L_1 = mR/(\pi f_1)$$

$$L^2 = \frac{(1 - m^2)R}{4\pi m f_1}$$

$$C = \frac{1 - m^2}{\pi f_1}$$

Shunt m-Derived Low-Pass Filter

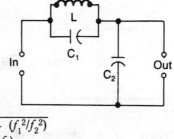

Fig. 6-3. Shunt m-derived low-pass filter.

$$m = \sqrt{1 - (f_1^2/f_2^2)}$$
$$L = mR/(\pi f_1)$$

$$C_1 = \frac{1 - m^2}{4\pi m f_1 R}$$

$$C_2 = m/(\pi f_2 R)$$

where f_1 = cutoff frequency
f_2 = frequency of remote cutoff
R = terminating resistance

Constant-k High-Pass Filter

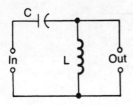

Fig. 6-4. Constant-k high-pass filter.

$$L = R/(4\pi f)$$
$$C = 1/(4\pi f R)$$

where f = cutoff frequency
R = terminating resistance

Series m-Derived High-Pass Filter

$$m = \sqrt{1 - (f_1^2 - f_2^2)}$$
$$L = R/(4\pi m f_2)$$
$$C_1 = 1/(4\pi m f_2 R)$$
$$C_2 = \frac{m}{(1 - m^2)\pi f_2 R}$$

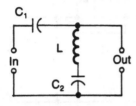

Fig. 6-5. Series m-derived high-pass filter.

where f_1 = frequency of maximum attenuation
       f_2 = frequency of maximum transmission
       R = terminating resistance

Shunt m-Derived High-Pass Filter

Fig. 6-6. Shunt m-derived high-pass filter.

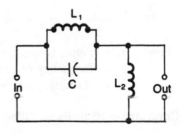

$$m = \sqrt{1 - (f_1^2/f_2^2)}$$

$$L_1 = \frac{mR}{(1 - m^2)\pi f_2}$$

$$L_2 = R/(4\pi m f_2)$$

$$C = 1/(4\pi m f_2 R)$$

where f_1 = frequency of maximum attenuation
       f_2 = frequency of maximum transmission
       R = terminating resistance

86

Constant-k Bandpass Filter

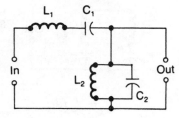

Fig. 6-7. Constant-k bandpass filter.

$$L_1 = \frac{R}{\pi(f_2 - f_1)}$$

$$L_2 = \frac{(f_2 - f_1)\,R}{4\pi f_1 f_2}$$

$$C_1 = \frac{f_2 - f_1}{4\pi f_1 f_2 R}$$

$$C_2 = \frac{1}{\pi(f_2 - f_1)R}$$

where f_1 = lower band-limit transmission frequency
f_2 = higher band-limit transmission frequency
R = terminating resistance

Series m-Derived Bandpass Filter

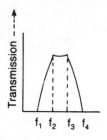

Fig. 6-8. Filter frequency response.

Note: f_1, f_2, f_3, and f_4 are the frequencies shown in the accompanying curve, and R is the terminating resistance.

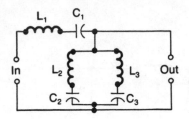

Fig. 6-9. Series m-derived bandpass filter.

$$f_1 = f_2 f_3/f_4$$

$$m = \frac{x}{1 - f_2 f_3/f_4^2}$$

$$x = \sqrt{\left(1 - \frac{f_2^2}{f_3^2}\right)\left(1 - \frac{f_3^2}{f_4^2}\right)}$$

$$y = \frac{(1 - m^2) f_2 f_3}{4f_1^2 x}\left(1 - \frac{f_1^2}{f_4^2}\right)$$

$$z = \left(\frac{1 - m^2}{4x}\right)\left(1 - \frac{f_1^2}{f_4^2}\right)$$

$$L_1 = \frac{mR}{\pi(f_3 - f_2)}$$

$$L_2 = \frac{zR}{\pi(f_3 - f_2)}$$

$$L_3 = \frac{yR}{\pi(f_3 - f_2)}$$

$$C_1 = \frac{f_3 - f_2}{4\pi f_2 f_3 mR}$$

$$C_2 = \frac{f_3 - f_2}{4\pi f_2 f_3 yR}$$

$$C_3 = \frac{f_3 - f_2}{4\pi f_2 f_3 zR}$$

Shunt m-Derived Bandpass Filter

In this circuit, f_1, f_2, f_3, f_4, m, x, y, z, and R are the same as for the series-type filter in the preceding section.

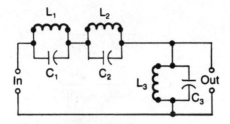

Fig. 6-10. Shunt m-derived bandpass filter.

$$L_1 = \frac{(f_3 - f_2)R}{4\pi f_2 f_3 z}$$

$$L_2 = \frac{(f_3 - f_2)R}{4\pi f_2 f_3 y}$$

$$L_3 = \frac{(f_3 - f_2)R}{4\pi f_2 f_3 m}$$

$$C_1 = \frac{y}{\pi(f_3 - f_2)R}$$

$$C_2 = \frac{y}{\pi(f_3 - f_2)R}$$

$$C_3 = \frac{m}{\pi(f_3 - f_2)R}$$

Constant-k Bandstop Filter

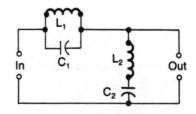

Fig. 6-11. Constant-k bandstop filter.

89

$$f_m = \sqrt{f_1 f_2}$$

$$L_1 = \frac{(f_2 - f_1)R}{\pi f_1 f_2}$$

$$L_2 = \frac{R}{4\pi(f_2 - f_1)}$$

$$C_1 = \frac{1}{4\pi(f_2 - f_1)R}$$

$$C_2 = \frac{f_2 - f_1}{\pi R f_1 f_2}$$

where f_1 = lower band-limit transmission frequency
f_2 = upper band-limit transmission frequency
f_m = center frequency in the suppressed band
R = terminating resistance

Series m-Derived Bandstop Filter

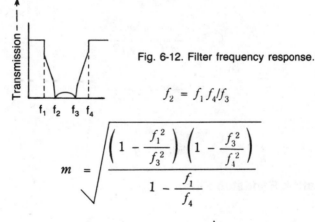

Fig. 6-12. Filter frequency response.

$$f_2 = f_1 f_4 / f_3$$

$$m = \sqrt{\frac{\left(1 - \dfrac{f_1^2}{f_3^2}\right)\left(1 - \dfrac{f_3^2}{f_4^2}\right)}{1 - \dfrac{f_1}{f_4}}}$$

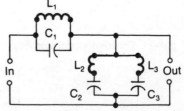

Fig. 6-13. Series m-derived band-stop filter.

$$x = \frac{1}{m} \left(1 + \frac{f_1 f_4}{f_3^{\,2}} \right)$$

$$y = \frac{1}{m} \left(1 + \frac{f_3^{\,2}}{f_1 f_4} \right)$$

$$L_1 = \frac{mR(f_4 - f_1)}{\pi f_1 f_4}$$

$$L_2 = \frac{R}{4\pi(f_4 - f_1)m}$$

$$L_3 = \frac{yR}{4\pi(f_4 - f_1)}$$

$$C_1 = \frac{1}{4\pi(f_4 - f_1)mR}$$

$$C_2 = \frac{f_4 - f_1}{\pi f_1 f_4 yR}$$

$$C_3 = \frac{f_4 - f_1}{\pi f_1 f_4 Rx}$$

where f_1, f_2, f_3, f_4 are the frequencies shown in the accompanying illustration, and R = terminating resistance.

Shunt m-Derived Bandstop Filter

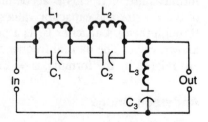

Fig. 6-14. Shunt m-derived bandstop filter.

In this circuit, $f_1, f_2, f_3, f_4, m, x,$ and y are the same as for the series-type filter in the preceding section.

$$L_1 = \frac{(f_4 - f_1)R}{\pi f_1 f_4 y}$$

$$L_2 = \frac{(f_4 - f_1)R}{\pi f_1 f_4 x}$$

$$L_3 = \frac{R}{4\pi(f_4 - f_1)m}$$

$$C_1 = \frac{x}{4\pi(f_4 - f_1)R}$$

$$C_2 = \frac{y}{4\pi(f_4 - f_1)R}$$

$$C_3 = \frac{m(f_4 - f_1)}{\pi f_1 f_4 R}$$

6.7 BRIDGE CIRCUITS AND NULL NETWORKS

Bridge circuits are especially important in the design of instruments used to measure component values, such as resistance and capacitance. Their primary characteristic is that the output signal voltage goes to zero (a *null*) when the various circuit values are brought into a specific relationship, at which time the bridge is said to be *balanced*. In addition to bridge circuits, another class of circuits, called null networks, also exhibits this important characteristic.

In the following circuits, components having a subscript "x" designate the position in the circuit in which components of unknown value are to be inserted for measurement. The circuits are brought into balance when the conditions imposed by the circuit formulas are met. This is accomplished by varying one or more of the resistor and capacitor values (shown with an arrow drawn through them). Capacitor C_S is a precision reference standard of known value. The natural frequency f and the angular velocity ω are related by the formula, $\omega = 2\pi f$.

Anderson Bridge

$$L_x = C_s[R_3(1 + R_2/R_4) + R_2]$$
$$R_x = R_1R_2/R_4$$

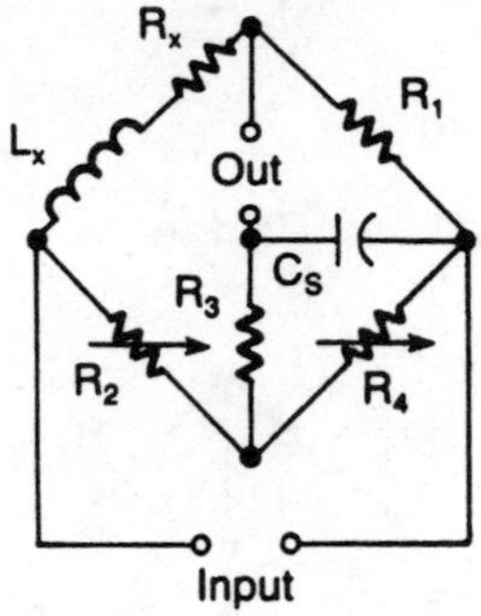

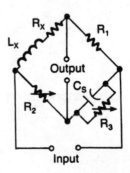

Fig. 6-15. Anderson bridge.

Hay Bridge

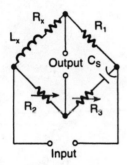

Fig. 6-16. Hay bridge.

$$L_x = C_s R_1 R_2$$

$$R_x = \frac{\omega^2 C_s^2 R_1 R_2 R_3}{1 + \omega^2 C_s^2 R_3^2}$$

$$Q = \omega L_x / R_x$$

Maxwell Bridge

Fig. 6-17. Maxwell bridge.

$$L_x = C_s R_1 R_2$$
$$R_x = R_1 R_2 / R_3$$

Owen Bridge

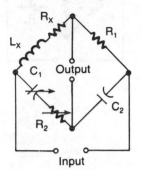

Fig. 6-18. Owen bridge.

$$L_x = C_2 R_1 R_2$$
$$R_x = R_1 C_2 / C_1$$

Schering Bridge

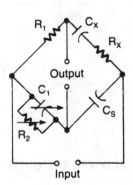

Fig. 6-19. Schering bridge.

$$C_x = C_s R_2 / R_1$$
$$R_x = R_1 C_1 / C_s$$

Wheatstone Bridge

$$R_x = R_2 R_3 / R_4$$

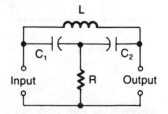

Fig. 6-20. Wheatstone bridge.

Wien Bridge

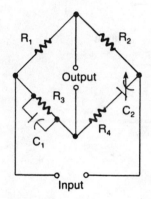

Fig. 6-21. Wien bridge.

$$C_1 = \sqrt{\frac{R_2R_3 - R_1R_4}{\omega^2 R_1 R_3{}^2 R_4}}$$

If $R_2 = 2R_1$, $C_1 = C_2$, and $R_3 = R_4$ at all settings, then

$$C_1 = 1/(2\pi f R_3)$$
$$f = 1/(2\pi R_3 C_1)$$

Bridged-T, LC Null Network

Fig. 6-22. Bridged-T LC null network.

When $C_1 = C_2$,

$$f_n = 1/(\pi\sqrt{2LC})$$

where f_n = null frequency

Bridged-T, RC Null Network

When $C_1 = C_2$,

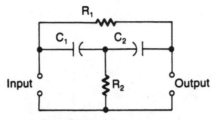

Fig. 6-23. Bridged-T RC null network.

$$f_n = \frac{1}{2\pi C\sqrt{R_1 R_2}}$$

where f_n = null frequency

$$R_1 = \frac{1}{R_2(2\pi f C)^2}$$

Parallel-T, RC Null Network

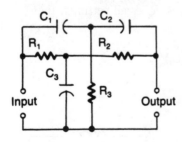

Fig. 6-24. Parallel-T RC null network.

When $C_1 = C_2 = \frac{1}{2}C_3$, and $R_1 = R_2 = 2R_3$,

$$f_n = 1/(2\pi R_1 C_1)$$

where f_n = null frequency

96

6.8 NOISE

Noise is a common problem in electronic circuits. The following formulas describe the thermal origin of noise and the basic parameters used in measuring it.

Thermal Noise (Johnson Noise or Resistor Noise)

$$N = kTB$$

where N = noise power (W)
 k = Boltzmann's constant (1.380×10^{-23} J/°K)
 T = absolute temperature (°K)
 B = bandwidth (Hz)

Thermal Noise Voltage

$$E_n = \sqrt{NR}$$

where E_n = noise voltage (V)
 R = resistance of noise source (ohms)
 N = noise power (from preceding formula)

Noise Figure or Noise Factor Ratio

$$F = \frac{(P_{si}/P_{ni})}{(P_{so}/P_{no})} = \frac{P_{no}}{G\,P_{ni}}$$

where F = noise figure (dimensionless)
 G = power gain of amplifier or system
 P_{ni} = noise input power (W)
 P_{no} = noise output power (W)
 P_{si} = signal input power (W)
 P_{so} = signal output power (W)

Noise Figure Ratio

$$NF = 10 \log_{10}(F)$$

where NF = noise figure ratio (decibels)
 F = noise figure

Noise Power

$$P_n = \frac{E^2}{(R_s + R_l)^2} \, R_l$$

where P_n = noise power (W)
$\quad\quad E$ = voltage of source (V RMS)
$\quad\quad R_s$ = internal resistance of source (ohms)
$\quad\quad R_l$ = load resistance (ohms)

Noise Bandwidth

$$\underline{B} = \int_0^x \frac{A_p(f)}{A_{pm}} \, df$$

where $\underline{B}$ = bandwidth
$\quad A_p(f)$ = power gain as a function of frequency
$\quad\quad A_{pm}$ = maximum value reached by A_p
$\quad\quad f$ = frequency (Hz)

This equation simplifies to $\underline{B} = A$, where A is the area under the curve of power gain vs. frequency; that is; A is the area of an equivalent rectangle having the same width and height (A_{pm}) displayed by the curve.

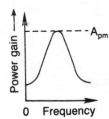

Fig. 6-25. Power gain versus noise frequency.

Signal-To-Nose Ratio

$$S/N = 10 \log_{10}(P_s/P_n)$$

where S/N = signal-to-noise ratio (decibels)
$\quad\quad P_n$ = noise power (W)
$\quad\quad P_s$ = signal power (W)

If the input and output signals are measured across the same or equal impedances, then

$$S/N = 20 \log_{10} (V_s/V_n)$$

where V_n = noise voltage (V)
V_s = signal voltage (V)

Chapter 7

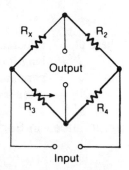

Magnetics Formulas

This chapter covers the basic relationships between the electrical, mechanical, and physical properties of magnetic circuits. These range from the behavior of magnetic circuits described by Ohm's law to the interaction of the windings in a transformer.

7.1 OHM'S LAW FOR MAGNETIC CIRCUITS

There is a great deal of similarity between magnetic circuits and electric circuits. The *reluctance* in a magnetic circuit is analogous to the resistance in an electric circuit. Similarly, *magnetomotive force* is analogous to voltage, and *lines of force* or *magnetic flux* are analogous to electric current. The relationships between these magnetic parameters may be stated as follows:

$$\phi = F/\mathcal{R}$$
$$R = F/\phi$$
$$F = \phi \mathcal{R}$$

where ϕ = lines of force (maxwells)
F = magnetomotive force (gilberts)
$\mathcal{R}$ = reluctance

7.2 MAGNETIC RELUCTANCE

The reluctance in a magnetic circuit obeys laws similar to those for resistance in an electric circuit. For more complex combina-

101

tions of series and parallel reluctances, simply refer to the chapter on resistance.

Reluctance in Terms of Geometry and Permeability

$$\mathcal{R} = l/\mu A$$

where $\mathcal{R}$ = reluctance
l = length of magnetic path (cm)
μ = permeability of core material
A = cross-sectional area of core (sq cm)

Reluctances in Series

$$\mathcal{R}_t = \mathcal{R}_1 + \mathcal{R}_2 + \mathcal{R}_3 + \ldots + \mathcal{R}_n$$

Reluctances in Parallel

$$\mathcal{R}_t = \cfrac{1}{\cfrac{1}{\mathcal{R}_1} + \cfrac{1}{\mathcal{R}_2} + \cfrac{1}{\mathcal{R}_3} + \ldots + \cfrac{1}{\mathcal{R}_n}}$$

7.3 BASIC MAGNETICS FORMULAS

When electric current flows through a coil or conductor in a magnetic circuit, various fields and forces are generated. The relationships between these parameters are summarized in the following formulas.

Flux Density

$$B = \phi/A$$

where B = flux density (gausses)
ϕ = lines of force (maxwells)
A = cross-sectional area of magnetic path (sq cm)

Permeability

$$\mu = B/H$$

where μ = permeability
B = flux density (gausses)
H = magnetizing force (oersteds)

Magnetomotive Force in Coil

$$F = 0.4\ \pi NI$$

where F = magnetomotive force (gilberts)
I = coil current (amperes)
N = number of turns in coil

Magnetizing Force

$$H = \frac{0.4\pi NI}{l}$$

where H = magnetizing force (oersteds)
I = coil current (amperes)
l = length of magnetic path (cm)
N = number of turns in coil

Force Acting on Current-Carrying Conductor Moving Perpendicular to Magnetic Field

$$F = BIl$$

where F = force (newtons)
B = flux density (webers/sq meter)
I = current (amperes)
l = length of conductor (meters)

7.4 INDUCED VOLTAGE IN MAGNETIC CIRCUITS

Voltage can be generated in a conductor by two different methods. The first is to vary the magnitude of the magnetic field surrounding the conductor. The second is to vary the position of the conductor within the magnetic field.

Voltage Generated in Conductor Cutting Across Magnetic Field

$$E = -BlV$$

where E = generated voltage (volts)
     B = flux density (webers/sq meter)
     l = length of conductor (meters)
     V = velocity of conductor (meters/sec)

Voltage Induced in Coil By Changing Flux

$$E = -N \frac{d\phi}{dt} \times 10^{-8}$$

where E = induced voltage (volts)
     N = number of turns in coil
     ϕ = flux linkage (maxwells)
  $d\phi/dt$ = derivative of flux with respect to time
     t = time (seconds)

7.5 POWER LOSS IN MAGNETIC CIRCUITS

Electric power is dissipated in a magnetic circuit through three different effects. The first loss is due to the ohmic resistance of the conductor carrying electric current. The second is hysteresis loss in the magnetic core due to the motion of the molecules. The third loss is due to eddy currents generated in the core, which dissipate energy through the electrical resistance of the core material.

Power Loss in Conductor

$$P_c + I^2R = E^2/R$$

where P_c = conductor power loss (watts)
     I = current flowing in conductor winding (amperes)
     E = voltage across conductor winding (volts)
     R = resistance of conductor winding (ohms)

Hysteresis Loss

$$P_h = 0.796 \, BH \times 10^{-8}$$

where P_h = hysteresis loss in core (watt-sec/cm^3/Hz)
   BH = area of measured hysteresis curve (flux density B in gausses vs magnetizing force H in oersteds)

Eddy Current Loss

$$P_e = K_e V f^2 t^2 B_{max}$$

where P_e = eddy current loss (watts)
B_{max} = maximum flux density (gauss)
K_e = constant depending upon core material and its insulation (for silicon steel, 4.1×10^{-12} is typical for K_e)
t = thickness of core (cm)
V = volume of core (cm³)

Total Power Loss

$$P_t = P_c + P_h + P_e$$

where P_t = total power loss (watts)
P_c = conductor loss (watts)
P_h = hysteresis loss (watts)
P_e = eddy current loss (watts)

7.6 TRANSFORMERS

An important use of magnetic circuits is the construction of transformers, which commonly have two or more interacting conductor windings. The basic transformer relationships are summarized in the following formulas.

Transformer Impedance Ratio

$$Z_p/Z_s = (N_p/N_s)^2$$

where N_p = number of primary turns
N_s = number of secondary turns
Z_p = primary impedance
Z_s = secondary impedance

Transformer Turns Ratio in Terms of Impedance Ratio

$$N_p/N_s = \sqrt{Z_p/Z_s}$$

where N_p = number of primary turns
N_s = number of secondary turns
Z_p = primary impedance
Z_s = secondary impedance

Transformer Voltage Ratio

$$E_p/E_s = N_p/N_s$$

where E_p = voltage across primary winding
E_s = voltage across secondary winding
N_p = number of primary turns
N_s = number of secondary turns

Transformer Current Ratio

$$I_s/I_p = N_s/N_p$$

where I_p = current in primary winding
I_s = current in secondary winding
N_p = number of primary turns
N_s = number of secondary turns

Chapter 8

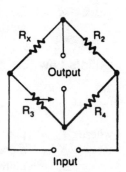

Vacuum
Tube Formulas

This chapter covers the basic formulas governing the behavior of vacuum tubes. This includes the characteristics of the vacuum tube diode, triode, and multigrid tubes. Circuit formulas are provided for determining gain, power, and component values.

8.1 BASIC LAWS OF VACUUM TUBES

The actual behavior of a vacuum tube depends upon much more than just the voltage applied between its plate and cathode. The basic physical laws governing the relationship between plate voltage and plate current are given by the following formulas.

Richardson—Dushman Equation

$$I = AT^2 \epsilon^{-b/T}$$

where I =   saturation current per sq cm of emitting surface (amperes)

      A =   a constant depending upon emitter material

      T =   temperature of emitter (°K)

      b =   a constant depending upon work function of emitter material

The constant b arises from the equation:

$$b = eW/k$$

where e = charge of electron (1.602×10^{-19} coulomb)
         k = Boltzmann's constant (1.38×10^{-23} watt-sec)
         W = work function of emitter material

Child's Law

$$I_p = KE_p^n$$

where I_p = plate current (A)
        K = a constant depending upon tube geometry
        E_p = plate voltage (V)
        n = a constant, approximately 3/2

Plate Current in Relation to Plate and Grid Voltage

$$I_p = G(\mu E_g + E_p)^n$$

where I_p = plate current (A)
        μ = amplification factor of tube
        E_g = grid voltage (V)
        E_p = plate voltage (V)
        n = a constant, approximately 3/2
        G = perveance of tube, a constant depending upon tube geometry

8.2 VACUUM TUBE DIODE

The relationship between plate voltage and plate current in a vacuum tube diode is provided by the following two formulas.

Diode Plate Current (three-halves power law)

$$I_p = G_d(E_p)^{3/2}$$

where I_p = plate current (A)
        E_p = plate voltage (V)

$$G_d = \text{perveance of diode, a constant depending upon tube geometry}$$

Perveance of Diode

$$G_d = 2.3 \times 10^{-6} \, A/d$$

where G_d = perveance
A = area of plate (sq cm)
d = plate—cathode separation (cm)

8.3 VACUUM TUBE TRIODE

Vacuum tube triodes and more complex multigrid tubes include a grid for controlling plate current. The following two formulas describe the basic operation of this grid element.

Triode Plate Current in Relation to Perveance

$$I_p = T_t [(E_p + \mu E_g)/(1 + \mu)]^{3/2}$$

where I_p = plate current
E_g = grid voltage
E_p = plate voltage
G_t = perveance of triode
μ = amplification factor of tube

Perveance of Triode

$$G_t = 2.3 \times 10^{-6} \, A/d$$

where G_t = perveance of triode
A = area of plate (sq cm)
d = grid—cathode separation (cm)

8.4 BASIC VACUUM-TUBE PARAMETERS

The operation of a vacuum tube is described using many different parameters, such as mu and transconductance. The basic parameters are summarized in this section.

Dc Plate Resistance

$$r_p = E_p/I_p$$

where r_p =   internal dc plate resistance (ohms)
        E_p =   dc plate—cathode voltage (V)
        I_p =   plate current (A)

Dynamic Plate Resistance

For constant E_g:

$$r_p = dE_p/dI_p$$

where r_p =   dynamic plate resistance (ohms)
        E_g =   grid voltage (V)
  dE_p/dI_p =   derivative of plate voltage with respect to plate current

Dc Screen Resistance

$$r_s = E_s/I_s$$

where r_s =   internal screen resistance of tube
        E_s =   screen voltage (V)
        I_s =   screen current (A)

Transconductance

For constant E_p:

$$g_m = dI_p/dE_g$$

where g_m =   transconductance (mhos)
        E_p =   plate voltage (V)
  dI_p/dE_g =   derivative of plate current with respect to grid voltage

Amplification Factor

For constant I_p:

$$\mu = dE_p/dE_g$$

where μ = amplification factor
dE_p/dE_g = derivative of plate voltage with respect to grid voltage
I_p = plate current (A)

Alternatively,

$$\mu = g_m r_p$$

where g_m = transconductance
r_p = internal plate resistance

Screen Amplification Factor

For constant I_s:

$$\mu_s = dE_s/dE_g$$

where μ_s = screen amplification factor
dE_s/dE_g = derivative of screen voltage with respect to grid voltage
I_s = screen current (A)

Output Resistance of Cathode Follower

$$R_o = R_k/(1 + g_m R_k)$$

where R_o = output resistance
g_m = transconductance of tube
R_k = resistance of cathode resistor (ohms)

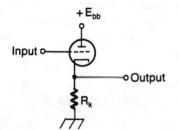

Fig. 8-1. Cathode-follower circuit.

111

Miller Effect

$$C_d = C_{gk} + C_{gp}(\mu + 1)$$

where C_d = dynamic input capacitance of tube
C_{gk} = static grid—cathode capacitance
C_{gp} = static grid—plate capacitance
μ = amplification factor

8.5 VACUUM TUBE CIRCUIT FORMULAS

The following formulas are useful in determining various voltages, currents, and resistances in vacuum-tube circuits.

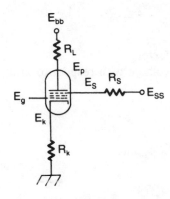

Fig. 8-2. Vacuum-tube circuit.

Required Dc Supply Voltage

$$E_{bb} = E_{pk} + E_k + E_L = E_{pk} + I_k R_n{}^k + I_p R_L$$

where E_{bb} = required supply voltage
E_k = required cathode—ground voltage
E_L = voltage across external plate load resistor
E_{pk} = required plate—cathode voltage
I_k = total cathode current
I_p = plate current
R_k = resistance of cathode resistor
R_L = resistance of external plate load resistor

Dc Plate—Cathode Voltage

$$E_p = E_{bb} - (I_p R_L + E_k)$$

where E_p = plate—cathode voltage
       E_{bb} = supply voltage
       E_k = cathode—ground voltage
       I_p = plate current
       R_L = resistance of external plate load resistor

Dc Screen Voltage

$$E_s = E_{ss} - I_s R_s$$

where E_s = screen—cathode voltage
       E_{ss} = screen supply voltage
       I_s = screen current
       R_s = resistance of external screen load resistor

Dc Screen Current

$$I_s = G(E_g + E_s/\mu_s)$$

where I_s = screen current
       E_g = control—grid voltage
       E_s = screen voltage
       G = perveance of tube
       μ_s = screen amplification factor

Required Plate Load Resistor

$$R_L = [E_{bb} - (E_p + E_k)]/I_p$$

where R_L = required resistance of external plate load resistor
      E_{bb} = available dc supply voltage
      E_k = required cathode—ground voltage
      E_p = required plate—cathode voltage
      I_p = required plate current

Required Cathode Resistor

$$R_k = (E_s - E_g)/I_k$$

where R_k = required resistance of cathode—ground resistor
       E_s = supply voltage
       E_g = required grid voltage
       I_k = total cathode current

Required Screen Resistor

$$R_s = (E_{ss} - E_s)/I_s$$

where R_s = required resistance of screen dropping resistor
       E_s = required screen voltage
       E_{ss} = screen supply voltage
       I_s = required screen current

8.6 AMPLIFICATION

A vacuum tube is basically a voltage amplifier, since the grid circuit normally draws a negligible amount of power. However, in high-power circuits a significant amount of grid current is drawn, and hence input power, so that such amplifiers may be considered to amplify power.

Voltage Amplification

$$A_v = g_m r_p R_L/(r_p + R_L) = \mu R_L/(r_p + R_L)$$

where A_v = voltage amplification
       g_m = transconductance
       μ = amplification factor
       r_p = internal plate resistance of tube
       R_L = external plate load resistance

If $r_p >> 10\ R_L$,

$$A_v = g_m R_L$$

Power Amplification

$$A_p = P_o/P_g$$

where A_p = power amplification
P_g = ac grid-input power
P_o = ac plate-output power

8.7 INPUT AND OUTPUT POWER FORMULAS

This section contains various formulas concerned with power utilization in vacuum tube circuits.

Heater (filament) Power

$$P_f = E_f I_f$$

where P_f = filament (heater) power
E_f = filament (heater) voltage
I_f = filament (heater) current

Dc Screen Power

$$P_s = E_s I_s$$

where P_s = screen power
E_s = screen voltage
I_s = screen current

Dc Plate Input Power

$$P_i = E_p I_p$$

where P_i = plate input power
E_p = plate voltage
I_p = plate current

Power Output

$$P_o = \frac{(E_{max} - E_{min})(I_{max} - I_{min})}{8}$$

115

where P_o = ac power output
E_{max} = maximum value of instantaneous plate voltage
E_{min} = minimum value of instantaneous plate voltage
I_{max} = maximum value of instantaneous plate current
I_{min} = minimum value of instantaneous plate current

Plate Power Dissipation

$$P_d = P_i - P_o$$

where P_d = plate power dissipation
P_i = dc plate power input
P_o = ac plate power output

Plate Efficiency

$$\eta_p = 100 \ (P_o/P_i)$$

where η_p = plate efficiency (%)
P_i = dc plate power input
P_o = ac plate power output

For class-A operation:

$$\eta_p = \frac{100 \ (E_{max} - E_{min})(I_{max} - I_{min})}{8E_pI_p}$$

where η_p = plate efficiency (%)
E_{max} = maximum value of instantaneous plate voltage
E_{min} = minimum value of instantaneous plate voltage
E_p = steady dc plate voltage
I_{max} = maximum value of instantaneous plate current
I_{min} = minimum value of instantaneous plate current
I_p = steady dc plate current

Power Sensitivity

$$PS = P_o/E_g$$

where PS = power sensitivity
E_g = ac grid driving voltage
P_o = ac power output

Chapter 9

Semiconductor Formulas

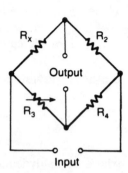

This chapter includes the most commonly encountered semiconductor formulas. These range from the basic physics of the semiconductor diode to field-effect transistor network formulas.

9.1 SEMICONDUCTOR DIODE

The basic formulas describing the behavior properties of the semiconductor diode are presented in this section. As with the vacuum tube diode, the semiconductor diode also exhibits a nonlinear relationship between voltage and current.

Diode Forward Current

$$I_F = I_s(\epsilon^{qV/kT} - 1)$$

where I_F = forward current
I_s = reverse saturation current
q = electron charge (1.602×10^{-19} coulomb)
V = forward voltage
k = Boltzmann's constant (1.380×10^{-23} J/°K)
T = temperature (degrees K)

Static Resistance of Diode or Rectifier

$$r_d = \frac{E_d}{I_d}$$

where r_d = static internal resistance of diode
E_d = dc voltage drop across diode
I_d = dc current through diode

Dynamic Resistance of Diode or Rectifier

$$r_d = \frac{de_d}{di_d}$$

where r_d = dynamic resistance of diode
e = diode voltage drop
i_d = diode current

Diode Rectification Efficiency

$$\eta = 100 \ (E_{DC}/E_{AC})$$

where η = efficiency (%)
E_{DC} = DC output voltage of diode
E_{AC} = peak value of AC input voltage = 1.414 E_{rms}

9.2 JUNCTION TRANSISTOR

The junction bipolar transistor is the most familiar type. The basic formulas for the transistor are presented in this section. Resistance and hybrid parameters are given in later sections.

Transistor Base Resistance

For constant V_c:

$$r_b = \frac{dV_b}{dI_b}$$

where r_b = base resistance
I_b = base current
V_b = base voltage

118

Transistor Emitter Diffusion Resistance

For constant V_c:

$$r_e = \frac{dV_e}{dI_e}$$

where r_e = emitter diffusion resistance
$\quad I_e$ = emitter current
$\quad V_c$ = collector voltage
$\quad V_e$ = emitter voltage

Transistor Collector Resistance

For constant I_e:

$$r_c = \frac{dV_c}{dI_c}$$

where r_c = collector resistance
$\quad I_c$ = collector current
$\quad I_e$ = emitter current
$\quad V_c$ = collector voltage

Transistor Emitter Feedback Conductance

For constant V_e:

$$g_{ec} = \frac{dI_e}{dV_c}$$

where g_{ec} = emitter feedback conductance
$\quad I_e$ = emitter current
$\quad V_c$ = collector voltage
$\quad V_e$ = emitter voltage

Transistor Amplification Factor (Alpha)

For constant V_c:

$$\alpha = \frac{dI_c}{dI_e}$$

where α = current amplification factor
$\quad\quad I_c$ = collector current
$\quad\quad I_e$ = emitter current
$\quad\quad V_c$ = collector voltage

Transistor Current Amplification Factor (Beta)

For constant V_c:

$$\beta = \frac{dI_c}{dI_b}$$

where β = current amplification factor
$\quad\quad I_b$ = base current
$\quad\quad I_c$ = collector current
$\quad\quad V_c$ = collector voltage

Transistor Alpha-to-Beta Conversion

$$\beta = \alpha/(1 - \alpha)$$

Transistor Beta-to-Alpha Conversion

$$\alpha = \beta/(\beta + 1)$$

Transistor Stability Factor

$$S = \frac{dI_c}{dI_{co}}$$

where S = stability factor
$\quad\quad I_c$ = collector current
$\quad\quad I_{co}$ = collector leakage current

9.3 TRANSISTOR RESISTANCE PARAMETERS

The resistance coefficients given in the following topics refer to the transistor equations:

$$V_{in} = R_{11}I_{in} + R_{12}I_{out}$$
$$V_{out} = R_{21}I_{in} + R_{22}I_{out}$$

120

where V_{in} = input voltage (V)
V_{out} = output voltage (V)
I_{in} = input current (A)
I_{out} = output current (A)

Transistor Common-Base Resistance Parameters

Input resistance:

$$R_{11} = r_e + r_b$$

Reverse transfer resistance:

$$R_{12} = r_b$$

Forward transfer resistance:

$$R_{21} = r_b + r_m$$

Output resistance:

$$R_{22} = r_c + r_b$$

where r_b = base resistance
r_c = collector resistance
r_e = emitter resistance
$r_m = \alpha r_c$
α = current amplification factor

Transistor Common-Emitter Resistance Parameters

Input resistance:

$$R_{11} = r_e + r_b$$

Reverse transfer resistance:

$$R_{12} = r_e$$

Forward transfer resistance:

$$R_{21} = r_e - r_m$$

Output resistance:

$$R_{22} = r_c + r_e - r_m$$

where r_b = base resistance
          r_c = collector resistance
          r_e = emitter resistance
          r_m = αr_c
          α = current amplification factor

Transistor Common-Collector Resistance Parameters

Input resistance:

$$R_{11} = r_b + r_c$$

Reverse transfer resistance:

$$R_{12} = r_c - r_m$$

Forward transfer resistance:

$$R_{21} = r_c(1 - \alpha)$$

Output resistance:

$$R_{22} = r_e + r_c - r_m$$

where r_b = base resistance
          r_c = collector resistance
          r_e = emitter resistance
          r_m = αr_c
          α = current amplification factor

9.4 TRANSISTOR HYBRID PARAMETERS

The hybrid coefficients in the following topics refer to the hybrid transistor equations:

$$V_{in} = h_i I_{in} + h_r V_{out}$$
$$I_{out} = h_f I_{in} + h_o V_{out}$$

where V_{in} = input voltage (V)
V_{out} = output voltage (V)
I_{in} = input current (A)
I_{out} = output current (A)

Transistor Common-Base Hybrid Parameters

Input resistance for constant V_{cb}:

$$h_{ib} = \frac{dV_{eb}}{dI_e}$$

Output conductance for constant I_e:

$$h_{ob} = \frac{dI_c}{dV_{cb}}$$

Forward transfer for constant V_{cb}:

$$h_{fb} = \frac{dI_c}{dI_e}$$

Reverse transfer for constant I_e:

$$h_{rb} = \frac{dV_{eb}}{dV_{cb}}$$

where I_c = collector current
I_e = emitter current
V_{cb} = collector-base voltage
V_{eb} = emitter-base voltage

Transistor Common-Emitter Hybrid Parameters

Input resistance for constant V_{ce}:

$$h_{ie} = \frac{dV_{be}}{dI_b}$$

Output conductance for constant I_b:

$$h_{oe} = \frac{dI_c}{dV_{ce}}$$

Forward transfer for constant V_{ce}:

$$h_{fe} = \frac{dI_c}{dI_b}$$

Reverse transfer for constant I_b:

$$h_{re} = \frac{dV_{be}}{dV_{ce}}$$

where I_b = base current
I_c = collector current
V_{be} = base-emitter voltage
V_{ce} = collector-emitter voltage

Transistor Common-Collector Hybrid Parameters

Input resistance for constant V_{ec}:

$$h_{ic} = \frac{dI_e}{dV_{ec}}$$

Output conductance for constant I_b:

$$h_{oc} = \frac{dI_e}{dV_{ec}}$$

Forward transfer for constant V_{ec}:

$$h_{fc} = \frac{dV_{bc}}{dI_b}$$

Reverse transfer for constant I_b:

$$h_{rc} = \frac{dV_{bc}}{dV_{ec}}$$

where I_b = base current
I_e = emitter current
V_{bc} = base-collector voltage
V_{ec} = emitter-collector voltage

9.5 FIELD-EFFECT TRANSISTOR (FET) FORMULAS

The following formulas are helpful when working with field-effect transistor circuits.

FET Transconductance

$$g_{fs} = 1000 \ \frac{dI_D}{dV_G}$$

where g_{fs} = common-source forward transconductance (μmhos)

I_D = drain current (mA)

V_G = gate voltage (volts)

FET Common-Source Voltage Gain (Unpassed Source Resistor)

$$A_v = \frac{g_{fs}R_L}{1 + g_{fs}R_s}$$

where A_v = voltage gain

g_{fs} = forward transconductance

R_L = resistance of external drain resistor

R_s = resistance of external source resistor

FET Common-Source Voltage Gain (Bypassed Source Resistor)

$$A_v = g_{fs}R_L$$

where A_v = voltage amplification

g_{fs} = forward transconductance

R_s = resistance of external source resistor

FET Source-Follower Formulas

Voltage gain:

$$A_v = \frac{g_{fs}R_s}{1 + g_{fs}R_s}$$

where A_v = voltage gain

g_{fs} = forward transconductance

R_L = resistance of external drain resistor

125

Output impedance:

$$Z_o = R_s/(1 + g_{fs}R_s)$$

where Z_o = output impedance
g_{fs} = forward transconductance
R_s = resistance of external source resistor

Chapter 10

Formulas for Antennas and Transmission Lines

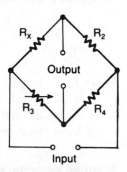

The formulas in this chapter cover basic relationships between frequency and wavelength, antenna dimensions and characteristics, matching networks, and transmission lines.

10.1 FREQUENCY, WAVELENGTH, AND PERIOD

The formulas which follow show the basic relationships between frequency, wavelength, and period. In addition, several topics are included for converting frequency to wavelength, and vice versa, using familiar units of length.

Frequency Units

$$
\begin{aligned}
\text{Hz} &= \text{hertz (one cycle per second)} \\
\text{kHz} &= \text{kilohertz } (10^3 \text{ Hz}) \\
\text{MHz} &= \text{megahertz } (10^6 \text{ Hz}) \\
\text{GHz} &= \text{gigahertz } (10^9 \text{ Hz}) \\
\text{THz} &= \text{terahertz } (10^{12} \text{ Hz})
\end{aligned}
$$

Frequency and Angular Velocity

$$
\begin{aligned}
f &= \omega/2\pi \\
\omega &= 2\pi f
\end{aligned}
$$

127

where f = frequency (Hz)
 ω = angular velocity (radians/second)
 π = 3.14159

Frequency and Wavelength

$$f = c/\lambda \cong (3 \times 10^8) / \lambda$$
$$\lambda = c/f \cong (3 \times 10^8)/f$$

where f = frequency (Hz)
 c = speed of light in a vacuum (2.99793 $\times$ 10^8 meters/second)
 λ = wavelength (meters)

Angular Velocity and Wavelength

$$\omega = 2\pi c/\lambda \cong (1.88 \times 10^9)/\lambda$$
$$\lambda = 2\pi c/\omega \cong (1.88 \times 10^9)/\omega$$

where ω = angular velocity (radians/second)
 λ = wavelength (meters)
 c = speed of light in a vacuum (2.99793 $\times$ 10^8 meters/second)
 π = 3.14159

Period and Frequency

$$T = 1/f = 2\pi/\omega$$

where T = time period required for one cycle (sec)
 f = frequency (Hz)
 ω = angular velocity (rad/sec)

Time and Wavelength

$$T = \lambda/c \cong / (3 \times 10^8)$$

where T = time period required for the wave to move a distance equal to one wavelength (sec)
 λ = wavelength (m)
 c = speed of light in a vacuum (2.99793 $\times$ 10^8 m/sec)

Converting Frequency and Wavelength

$$f_{Hz} = 300{,}000{,}000/\lambda_m = 984{,}000{,}000/\lambda_{ft} = 11{,}800{,}000{,}000/\lambda_{in}$$
$$f_{kHz} = 300{,}000/\lambda_m = 984{,}000/\lambda_{ft} = 11{,}800{,}000/\lambda_{in}$$
$$f_{MHz} = 300/\lambda_m = 984/\lambda = 11{,}800/\lambda_{in}$$
$$f_{GHz} = 0.3/\lambda_m = 0.984/\lambda_{ft} = 11.8/\lambda_{in}$$

where λ_m = wavelength in meters
λ_{ft} = wavelength in feet
λ_{in} = wavelength in inches

$$\lambda_m = 300{,}000{,}000/f_{Hz} = 300{,}000/f_{kHz} = 300/f_{MHz} = 0.3/f_{GHz}$$
$$\lambda_{ft} = 984{,}000{,}000/f_{Hz} = 984{,}000/f_{kHz} = 984/f_{MHz} = 0.984/f_{GHz}$$
$$\lambda_{in} = 11{,}800{,}000{,}000/f_{kHz} = 11{,}800/f_{MHz} = 11.9/f_{GHz}$$

where λ_m = wavelength in meters
λ_{ft} = wavelength in feet
λ_{in} = wavelength in inches

Formulas for One-Half Wavelength

$$l_m = 150/f_{MHz}$$
$$l_{ft} = 492/f_{MHz}$$
$$l_{in} = 5905/f_{MHz}$$

where l_m = one-half wavelength in meters
l_{ft} = one-half wavelength in feet
l_{in} = one-half wavelength in inches
f_{MHz} = frequency in megahertz

10.2 ANTENNA DIMENSIONS

An actual antenna is a little shorter than the predicted wavelength. The following formulas are derived from real antennas in which the actual antenna length is approximately 95% of theoretical length.

Physical Length of Half-Wave Antenna

$$l_m = 143/f_{MHz}$$

$$l_{in} = 561/f_{MHz}$$
$$l_{ft} = 468/f_{MHz}$$

where l_m = length in meters
l_{ft} = length in feet
l_{in} = length in inches
f_{MHz} = frequency in megahertz

Physical Length of Harmonic Antenna

$$l = 492(N - 0.05)/f$$

where l = length (ft)
N = number of half-waves on antenna
f = frequency (MHz)

Length of Long-Wire Antenna

$$l = 984(N - 0.025)/f$$

where l = wire length (ft)
N = number of full waves on wire
f = frequency (MHz)

Dimensions of Elements in Parasitic Beam Antenna

$$l_R = 475/f$$
$$l_d = 455/f$$
$$l_r = 500/f$$

where l_R = length of radiator (ft)
l_d = length of director (ft)
l_r = length of reflector (ft)
f = frequency (MHz)

10.3 ANTENNA CHARACTERISTICS

The following formulas describe the basic electrical characteristics of antennas.

Gain of Directive Antenna, Such as a Parabolic Reflector

$$A_p = 8A/\lambda^2$$

where A_p = power gain
  A = area of aperture (sq cm)
  λ = wavelength (cm)

Radiation Resistance of Antenna

$$R_r = P_r/I^2$$

where R_r = radiation resistance
  P_r = radiated power (watts)
  I = current flowing in a resistor inserted at point of interest in antenna, which dissipates, same energy as that radiated by antenna

Antenna Efficiency

$$\eta_a = 100\, R_o/(R_o + R_x)$$

where η_a = antenna efficiency (%)
  R_o = radiation resistance
  R_x = loss resistance of antenna

10.4 TRANSMISSION LINES

Transmission lines are used to carry signals between two points, usually between a transmitter and antenna or between an antenna and a receiver.

General Impedance of Transmission Line

$$Z_o = \sqrt{L/C}$$

where Z_o = impedance of line (ohms)
  C = capacitance of line (pF/ft)
  L = inductance of line (μH/ft)

Characteristic Impedance of Two-Wire Line

$$Z_o = 276 \log_{10}(2\ S/d)$$

where Z_o = characteristic impedance
d = diameter of wire
S = spacing between centers of wires

(S and d are in same units.)

Impedance of Line an Odd Number of Quarter-Waves Long

$$Z_o = \sqrt{Z_i Z_L}$$

where Z_o = line impedance
Z_i = input impedance
Z_L = load impedance

Characteristic Impedance of Coaxial Line (Air-Insulated)

$$Z_o = 138 \log_{10}(b/a)$$

where Z_o = characteristic impedance
a = outside diameter of inner conductor
b = inside diameter of outer conductor

(a and b are in same units.)

10.5 MATCHING SECTIONS

Matching sections are used to connect a transmission line of one impedance to an antenna of different impedance.

Impedance of Q-Matching Section

$$Z_Q = \sqrt{Z_a Z_L}$$

where Z_Q = impedance of Q-matching section
Z_a = antenna impedance (resistance)
Z_L = line impedance

Delta Matching Transformer Dimensions

For matching 600-ohm line to antenna.

$$A = 118/f, \text{for } f \le 30 \text{ MHz}$$
$$A = 113/f, \text{ for } f > 30 \text{ MHz}$$
$$B = 148/f$$

where A = width of delta (in.)
B = height of delta (in.)
f = frequency (MHz)

T-Matching Section for 600-Ohm Line to Dipole Antenna

$$A = 180.5/f$$
$$B = 114/f$$

where A = length of horizontal T member (in.)
B = length of vertical T member (in.)
f = frequency (MHz)

10.6 STANDING-WAVE RATIO AND REFLECTION COEFFICIENT

The standing-wave ratio is a measure of the mismatch in transmission line impedances. For a perfect impedance match, the standing-wave ratio would be unity. When the impedances are not perfectly matched, reflections occur at the terminal junctions, setting up standing waves.

Definition of Standing-Wave Ratio (SWR)

$$\text{SWR} = R_L/Z_o$$

where SWR = standing-wave ratio
R_L = load resistance
Z_o = line impedance

When $R_L > Z_o$, $\text{SWR} = Z_o/R_L$

Reflection Coefficient in Terms of Standing-Wave Ratio

$$k = (SWR - 1) / SWR$$

where k = reflection coefficient
SWR = standing-wave ratio

Reflection Coefficient in Terms of Voltages

$$k = E_r / E_f$$

where k = reflection coefficient
E_f = forward voltage
E_r = reflected voltage

Reflection Coefficient in Terms of Load Resistance

$$k = (R_L - Z_0)/(R_L + Z_0)$$

where k = reflection factor
R_L = load resistance
Z_0 = characteristic impedance of line

10.7 LINE LOSS

Generally, line loss is caused by dielectric and conductor heating. This loss increases as the conductor size and spacing are decreased. Dielectric loss depends on the kind of dielectric material used, and also on the portion of the electromagnetic field that passes through the dielectric. Line loss increases with frequency, and is measured in decibels per unit length (per foot or per hundred feet). Some types of transmission lines along with their loss per foot at various frequencies are shown below.

Loss per Foot for Transmission Lines

Type of Line	1 MHz	10 MHz	100 MHz
Open wire	0.0005	0.001	0.005
TV "Twin Lead"	0.001	0.005	0.015

Type of Line	1 MHz	10 MHz	100 MHz
RG-8/U Coaxial	0.0015	0.006	0.02
RG-59/U Coaxial	0.003	0.01	0.04
RG-58/U Coaxial	0.003	0.014	0.05

The loss is increased by the presence of standing waves on the line. This is because the standing waves cause generally higher currents in the line conductors and higher voltages across the line dielectric.

Standing-Wave-Ratio Loss

The standing-wave-ratio loss is an additional loss that results from a mismatch between a feed line and its load. The standing-wave-ratio loss, or SWR loss, is a function of the matched-line loss and the SWR. The table shows SWR loss as a function of line loss when matched and the SWR.

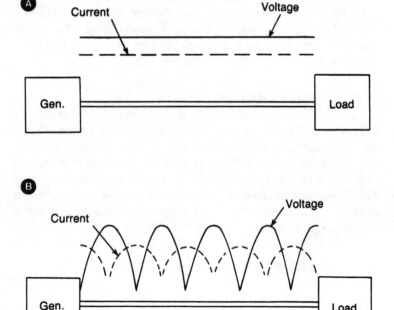

Fig. 10-1. Standing-wave ratio on a transmission line. At A, a "flat" line (perfect match). At B, a mismatched line.

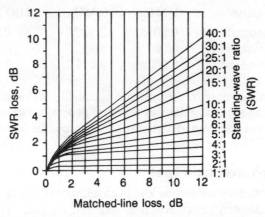

Fig. 10-2. Standing-wave-ratio loss can be determined with this chart. Find the matched-line loss on the horizontal axis, and the standing-wave ratio (as determined at the antenna feed point) among the family of curves. The corresponding point on the vertical axis represents the loss caused by standing waves.

The SWR must be measured from the feed point, where the line joins its load. It is assumed that the line loss under perfectly matched conditions is already known. The preceding table should give accurate values for new lines. Old lines may have dielectric contamination, which will increase the loss.

10.8 VELOCITY FACTOR

The velocity factor is defined as that percentage of the speed of light that an electromagnetic field travels along a transmission line. In an antenna, or in a single-wire or open-wire feed line, the velocity factor v is about 95 percent (0.95). For lines containing a dielectric other than air, the velocity factor is considerably less. This factor must be taken into account when calculating the physical length of a transmission-line stub for impedance matching, or when the electrical length of a line must be known.

Velocity Factors for Transmission Lines

Line Type or Manufacturer's No.	Velocity Factor, Percent
Coaxial cable, RG-58/U, solid dielectric	66

Fig. 10-3. Velocity factors of various types of transmission lines.

136

Line Type or Manufacturer's No.	Velocity Factor, Percent
Coaxial cable, RG-59/U, solid dielectric	66
Coaxial cable, RG-8/U, solid dielectric	66
Coaxial cable, RG-58/U, foam dielectric	75 - 85
Coaxial cable, RG-59/U, foam dielectric	75 - 85
Coaxial cable, RG-8/U, foam dielectric	75 - 85
Twin-lead, 75-ohm, solid dielectric	70 - 75
Twin-lead, 300-ohm, solid dielectric	80 - 85
Twin-lead, 300-ohm, foam dielectric	85 - 90
Open-wire with plastic spacers, 300-ohm	90 - 95
Open-wire with plastic spacers, 450-ohm	90 - 95
Open-wire, homemade, 600-ohm	95

Fig. 10-3. Continued.

10.9 LONGWIRE ANTENNAS

Longwire antennas are straight conductors, fed at either a current or voltage loop, designed so that they produce directional gain. The longwire is generally fed either at one end or at an odd multiple of ¼ wavelength from one end. The advantage of end feed is that the antenna can be tuned for operation at any wavelength.

The gain of a longwire antenna is a function of its length in electrical wavelengths. The length W in wavelengths is given by the following formulas.

Length of Longwire in Wavelengths

$$W = Lf/984$$

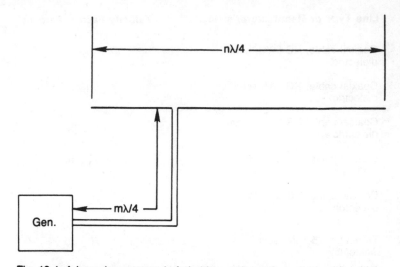

Fig. 10-4. A longwire antenna is fed either at the end, or at an odd multiple of ¼ wavelength from one end, as shown here (*m* is an odd integer and *n* may be any integer).

where L is the length in feet and f is the frequency in megahertz. For measurements made in meters,

$$W = Lf/300$$

where L is the antenna length.

Longwire Gain

The gain of a longwire antenna, in its favored directions, increases with increasing length. The graph below shows the approximate gain of a longwire antenna with respect to a half-wave dipole in free space.

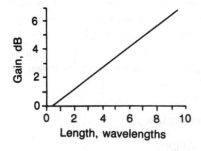

Fig. 10-5. Longwire gain, relative to a half-wave dipole in free space, as a function of the wire length in wavelengths.

Chapter 11

Impedance: A Complex Entity

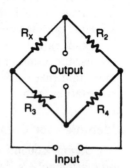

Impedance may be thought of as having two components: a resistance and a reactance. Resistance is the simple characteristic of a circuit or component to oppose the flow of current. *Resistance* is measured in ohms and does not depend on whether the current is direct or alternating, nor on the frequency if the current is alternating. *Reactance*, also measured in ohms, depends on the frequency of alternating current. In the case of direct current, reactance is identical to resistance.

In electronics engineering, impedance is generally represented as a *complex* quantity. This does not mean that it is complicated, but that it consists of two vector components, a real-number resistance and a reactance that is denoted by an imaginary number—a multiple of the operator j, where $j = \sqrt{-1}$.

11.1 THE j OPERATOR

Mathematicians usually represent the value $\sqrt{-1}$ by the small letter i, which stands for "imaginary." Electronics people use the lowercase letter j. Multiples of j are written nj or jn, where n is some real number. The value of n may be positive, negative, or zero. If n is negative, we have a capacitive reactance; if n is positive, an inductive reactance; if $n = 0$ there is no reactance. A complex impedance is thus written in the form as follows:

Complex Impedance in General

$$Z = R + jX,$$

where R is the resistance component (in ohms, kilohms, or megohms) and X is the real-number reactive component (in ohms, kilohms, or megohms, negative if a capacitance and positive if an inductance). If there is no reactance we write $R + j0$. If there is no resistance but only a pure capacitance we write $0 + jX$. It is important that both components be included in the complete representation even if one, or both, are zero.

Identities for Complex Numbers

Complex quantities add, subtract, multiply and divide just like ordinary numbers, but we must keep in mind that $j^2 = -1$. If R_1, R_2, X_1, and X_2 are the resistance and capacitance, which are real-number values, then

$$(R_1 + jX_1) + (R_2 + jX_2) = (R_1 + R_2) + j(X_1 + X_2)$$
$$(R_1 + jX_1) - (R_2 + jX_2) = (R_1 - R_2) + j(X_1 - X_2)$$
$$-jX_1 = j(-X_1)$$
$$(R_1 + jX_1)(R_2 + jX_2) = R_1R_2 + jR_1X_2 + jR_2X_1 - X_1X_2$$
$$(R_1 + jX_1)(R_1 - jX_1) = R_1^2 + X_1^2$$
$$(R_1 + jX_1) + (R_1 - jX_1) = 2R_1 + j0$$

The value $R_1 - jX_1$ is called the *conjugate* impedance of $R_1 + jX_1$.

$$1/(R_1 + jX_1) = (R_1 - jX_1)/(R_1^2 + X_1^2)$$

This is called the *reciprocal* impedance of $R_1 + jX_1$.

$$|R_1 + jX_1| = \sqrt{R_1^2 + X_1^2}$$

This is called the *absolute value* impedance for $R_1 + jX_1$. It is also occasionally called the impedance, although this is an oversimplification because, for a given Z, there are infinitely many combinations of R_1 and X_1 that will yield Z.

11.2 THE COMPLEX-IMPEDANCE COORDINATE PLANE

Complex numbers can be represented on a coordinate plane, in the same way as ordered number pairs as described in Chapter

1. Generally, the resistance is shown on the abscissa and the reactance on the ordinate, as shown below. The abscissa is given values R and the ordinate is given values jX. In practice, X can take any real-number value but R is positive or zero. (Negative resistances are shown in the figure since cases occasionally arise where a resistance may be considered negative; for example, a current source.) The axes are numbered conveniently so that the displayed impedances are easy to see. Usually the abscissa scale is the same as the ordinate scale.

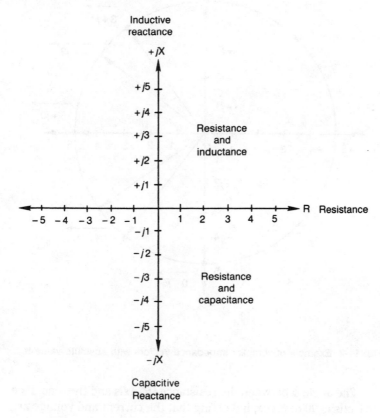

Fig. 11-1. The complex impedance plane.

An impedance is represented by a point on this plane. Each point corresponds to a unique impedance $R + jX$. Pure resistances lie on the horizontal axis and pure reactances along the vertical axis. The point $R + jX$ is often denoted as a vector quantity. The length of the vector is the absolute-value impedance. Clearly, an infinite

141

number of complex impedances can be obtained from a given absolute-value impedance by rotating the vector as shown. The illustration denotes several possible impedances having an absolute value of 5 ohms.

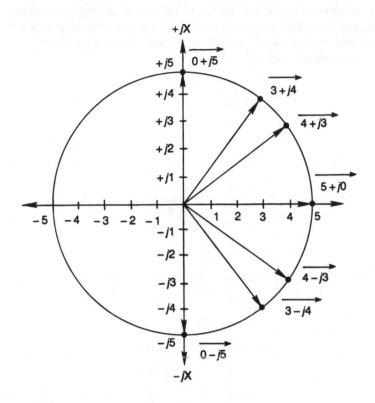

Fig. 11-2. Example of complex impedance vectors with absolute value 5.

The angle ϕ between the resistance (R) axis and the reactance (jX) axis is 90 degrees, indicating that the current and voltage are 90 degrees out of phase in a pure inductance or a pure capacitance. If the complex impedance contains both resistance and reactance, the angle ϕ will be between 0 and +90 degrees for an inductance and between 0 and −90 degrees for a capacitance. Two examples are shown.

The angle ϕ can be determined by the formula as follows:

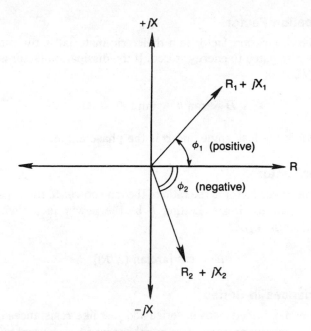

Fig. 11-3. Complex impedances with positive and negative angle.

Phase Angle

$$\phi = \arctan (X/R)$$

where X is the reactance (positive if inductive, negative if capacitive) and R is the resistance. The phase angle is the value by which the voltage leads the current, or by which the current lags the voltage. Angles between -90 and 0 degrees are sometimes considered to range from $+270$ to $+360$ degrees, where $+270 = -90$ and $+360 = 0$.

Loss Angle

$$\theta = 90 - \phi$$

where θ is the loss angle and ϕ is the phase angle. In a perfect inductance the loss angle is 0. In a perfect capacitance it is 180. In some cases we consider only the absolute value of ϕ in calculating the loss angle, since in practice, for any given loss angle θ, θ is equivalent to $180 - \theta$.

Dissipation Factor

The dissipation factor in a dielectric material is the ratio of energy dissipated to energy stored. If the dissipation factor is given by D, then

$$D = \tan \theta = \tan (90 - \phi)$$

where θ is the loss angle and ϕ is the phase angle.

Power Factor

The power factor is the ratio of the true power to the apparent power in an ac circuit. Letting F be the power factor with an impedance $R + jX$,

$$F = \cos [\arctan (X/R)]$$

Impedances in Series

Complex impedances in series add just like resistances in series. It is necessary, again, to remember that $j^2 = -1$ and to keep track of signs when multiplying. Given $Z_1 = R_1 + jX_1$ and $Z_2 = R_2 + jX_2$,

$$Z_1 + Z_2 = (R_1 + jX_1) + (R_2 + jX_2) = R_1 + R_2 + j(X_1 + X_2)$$

In series, impedances of equal and opposite value cancel each other. This leaves only the sum of the resistance components. This

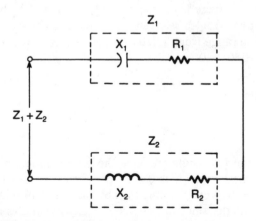

Fig. 11-4. Addition of impedances in series.

condition is called *resonance*. Resonance occurs in any series combination of inductance(s) and capacitance(s) at a particular frequency.

Impedances in Parallel

Impedances in parallel add just like resistances in parallel, if the complex representations are used. With the denotations as above, the parallel sum is

$$Z_1 + Z_2 \ldots + Z_n =$$

$$\frac{(R_1 + jX_1)(R_2 + jX_2)}{(R_1 + R_2 + j(X_1 + X_2) + \ldots + (R_n - jX_n)/(R_n^2 + X_n^2)]}$$

An alternative formula, useful for any number of impedances in parallel, is

$$Z_1 + Z_2 = 1/[(R_1 - jX_1)/(R_1^2 + X_1^2) + (R_2 - jX_2)/(R_2^2 + X_2^2)]$$

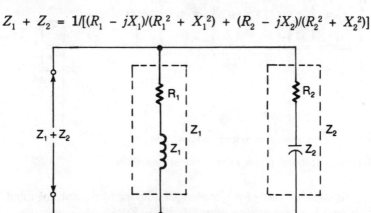

Fig. 11-5. Addition of impedances in parallel.

Characteristic Impedance

The characteristic impedance of a transmission line is not a true impedance, as such, but behaves that way provided the line is terminated in a purely resistive impedance of the right value. Characteristic impedances add in parallel exactly as do resistances in parallel. Thus, for two transmission lines in parallel, having equal length and having characteristic impedances Z_{01} and Z_{02}, the parallel characteristic impedance Z_0 is

$$Z_0 = Z_{01}Z_{02}/(Z_{01} + Z_{02}) \text{ or}$$

$$Z_0 = 1/(1/Z_{01} + 1/Z_{02})$$

The second formula can be used for any number of lines in parallel. The formulas assume there is no interaction between or among the lines.

Eliminating Reactance

Unless a transmission line is terminated in a pure resistance $R + j0$ equal to the line Z_0 in ohms, different complex impedances will appear at various points along the line. Assuming a lossless line, terminated in a pure resistance different from the line Z_0, resonances will occur at even multiples of ¼ wavelength from the terminated end. These resonances will show resistive values equal to R, as shown.

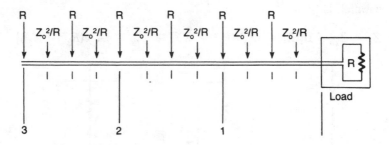

Fig. 11-6. Resonant points along a transmission line.

At odd multiples of ¼ wavelength from the feed point, the input will be a pure resistance equal to $R^* = Z_0^2/R$.

At intermediate distances from the feed point, reactance will be introduced. This is not desirable or useful in itself. However, if the terminating impedance does contain reactance, the introduced transmission-line reactance can be used to cancel the effect of the reactance at the terminating point. This introduced reactance is a function of the length of the line. Specific formulas are not given here since individual cases vary greatly. The design of a stub to eliminate reactance must be done on a case-by-case basis.

11.3 COMPLEX IMPEDANCES IN POLAR COORDINATES

The complex plane may be expressed in terms of polar coordinates as well as Cartesian coordinates. A pure resistance is indicated when the angle θ is zero. Pure reactances occur at $+90$ and -90 degrees (inductive and capacitive, respectively).

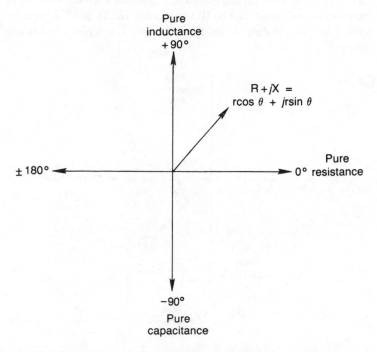

Fig. 11-7. Relationship of polar expression to Cartesian expression.

Absolute values are indicated by concentric circles, where the absolute value increases as the radius increases. The polar form is easier for expressing phase relationships between two impedance vectors, but does not directly denote the resistive and reactive components of a complex impedance. Given a complex impedance $R + jX$ and its vector equivalent $(\overrightarrow{\mathbf{r},\theta})$ in polar form, the following formulas apply.

Complex Impedance from Polar Vector

$$R + jX = [r \cos \theta + j(r \sin \theta)]$$

Polar Vector from Complex Impedance

$$(\overrightarrow{\mathbf{r},\theta}) = \{\sqrt{R^2 + X^2}, [\arctan(X/R)]\}$$

Here, it can be seen that any given absolute-value impedance will be represented by a vector $(\overrightarrow{r,\theta})$ where r is a constant and θ may vary from -90 through zero to $+90$. For a given value of θ, the ratio of the resistance to the reactance (R/X) or (X/R) is constant, but the absolute value may change. Examples of this are shown.

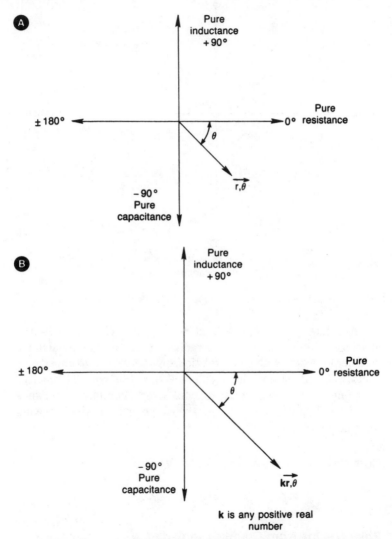

Fig. 11-8. At A, an impedance vector having an angle θ relative to the resistance axis. At B, θ is the same but the absolute value is changed by a factor of k, where k is a positive real number.

148

11.4 THE SMITH CHART

The *Smith chart* is a special form of coordinate system that is used for plotting complex impedances. The Smith chart is a curvilinear system; that is, the coordinate lines are not necessarily straight. Smith charts are especially useful in determining the resistance and reactance at the input end of a transmission line when the resistance and reactance at the feed point are known. The Smith chart is named after P.H. Smith, the inventory of the system.

On the Smith chart, resistance coordinates are represented by eccentric circles, mutually tangent at the bottom of the graph. The reactance coordinates are parts of circles having variable diameter and centering within the outer periphery. This allows any combination of resistance and reactance to be represented. The values assigned to the circles depend on the range of values to be shown on the chart (as is the case with any graph system). Usually, these values depend on the characteristic impedance of the transmission line. The illustration is an example of a Smith chart intended for analysis of feed systems in which the characteristic impedance of the line is 50 ohms ($Z_0 = 50$). The resistance line is such that 50 ohms appears at the center. In this particular case, the 50-ohm inductive reactance point is directly toward the right, and the 50-ohm capacitive reactance point is directly toward the left. Zero reactance is at the top and infinite reactance and resistance is at the bottom. These represent short and open circuits, respectively. We might use other values for the scale of reactance, depending on the situation, but this is not usually done since it distorts the scale of the Smith chart. The resistance values run in straight lines across the graph for "linear" scaling, as shown in the second illustration, so that the SWR circles (soon to be discussed) will appear truly circular.

Complex impedances appear as points on the Smith chart. Each point on the chart corresponds to one particular complex impedance, and vice-versa, so that there is a one-to-one correspondence. In this way, the Smith chart is similar to the complex coordinate plane, except for the fact that the Smith chart also shows infinite reactances and resistances. Negative resistances are not shown on the Smith chart.

Pure resistances, that is, impedances having the form $R + j0$, lie on the resistance line. The top of the line represents a short circuit and the bottom of the line represents an open circuit. Pure reactances, of the form $0 + jX$, where X may be any real number (including infinity in theory) lie on the perimeter of the circle, with

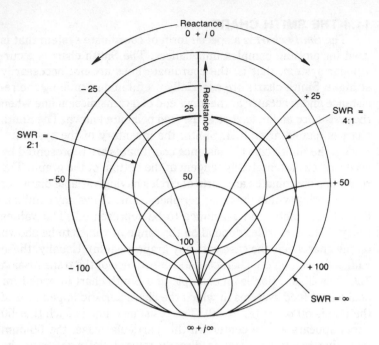

Fig. 11-9. Smith chart for $Z_0 = 50$ ohms.

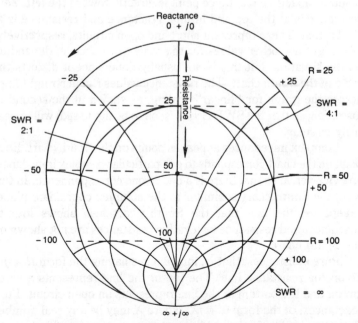

Fig. 11-10. Resistance lines (dotted) on the Smith chart are horizontal.

inductance to the right and capacitance to the left. Complex impedances of the form $R + jX$, where R is a positive real number and X may be any nonzero real number, lie within the circle. Several different complex points are shown here.

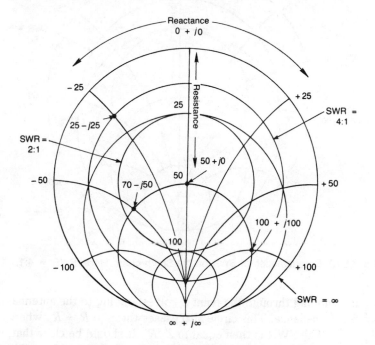

Fig. 11-11. Some complex impedance points on the Smith chart for $Z_0 = 50$ ohms.

The Smith chart can be used to determine the standing-wave ratio (SWR) on a transmission line, if the characteristic impedance of the line and the complex impedance of the antenna system are known. For determination of SWR, there are coordinates known as SWR circles. These circles are drawn on the Smith chart concentrically around the point on the resistance line marking the characteristic impedance of the transmission line. (For the SWR circles to be true circles, the Z_0 of the line must therefore be at the center of the Smith chart.) SWR circles for a 50-ohm line will intersect the resistance line at a certain point $R = R_x$, where $R_x < Z_0$ and Z_0/R_x is the SWR on the line.

An example of SWR determination is shown. The antenna system impedance is plotted as a point on the Smith chart. Then a compass can be used to draw the SWR circle, centered at $R =$

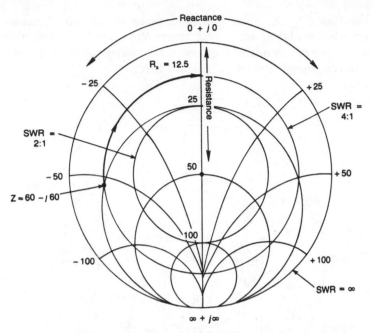

Fig. 11-12. Determination of SWR on the Smith chart. Here, SWR = 4:1.

Z_0 and running through the point Z corresponding to the antenna-system impedance. This circle also passes through $R = R_x$, where $R_x < Z_0$. The SWR is then equal to Z_0/R_x. It should be clear that, for any given SWR not 1:1, there exist infinitely many possible complex impedances that might result in that SWR. Also, note that the SWR cannot be 1:1 if reactance is present in the antenna system.

In a theoretically lossless line, the standing-wave ratio does not change as the measuring point is moved farther and farther from the feed point. In practice this ideal is not achieved; the line has loss. The complex impedance moves clockwise around the SWR circle from the point Z, with the angle varying in accordance with the number of electrical degrees that the measuring point is from the feed point.

The distance in electrical degrees is given by the following formula.

Distance from Feed Point in Degrees

$$d = 0.366fL/v$$

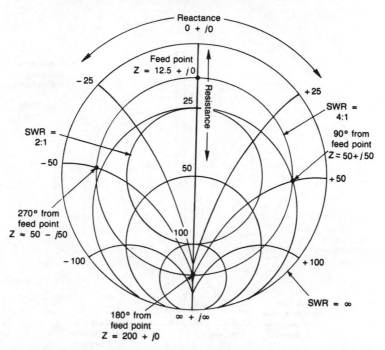

Fig. 11-13. Variation of complex impedance with distance from feed point when the antenna has impedance of 12.5 ohms purely resistive.

where d is the distance in electrical degrees, v is the velocity factor of the line, L is the distance from the feed point in feet, and f is the frequency in megahertz. For distances L in meters,

$$d = 1.20 \ fL/v$$

An example of this is shown in the illustration on the next page. Here, the antenna is a pure resistance of 100 ohms, so that the SWR is 2:1. The velocity factor of the 50-ohm coaxial line is 0.66. The complex impedances are shown for distances of 10, 50, 100, and 150 feet at a frequency f = 7.0 MHz. As an example, at L = 50 feet,

$$d = 0.366 \times 7.0 \times 50/0.66$$
$$= 194 \text{ degrees}$$

The angle is therefore measured clockwise from the ray starting at the center of the Smith chart and passing through R = 100. This can be done with a protractor, and the resulting point Z passes

153

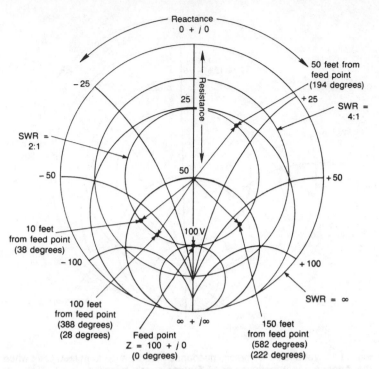

Fig. 11-14. A specific example of displacement of complex impedance vector with distance from feed point.

through the ray at 194 degrees and the 2:1 SWR circle. Note that angles greater than 360 degrees are considered between 0 and 360 degrees, simply by subtracting 360 as many times as necessary to obtain a number between 0 and 360. This occurs here for $L = 100$ and $L = 150$.

This technique can be used in order to determine the required amount of feed line necessary to provide a pure resistive input at the transmitter end of a line that is not perfectly matched. If the antenna is a pure resistance $R \neq Z_0$, then the input impedance will be resistive when the line is an integral multiple of ½ wavelength. If the antenna contains reactance, the point is first plotted on the Smith chart corresponding to the complex antenna impedance. Then the angle is measured between this point and the resistance axis. Actually this may be two angles, say θ and $180 - \theta$, as shown. In addition to these, we may use feed lines having electrical lengths of $\theta + 180\,n$, where n is any integer. The physical length of such a line is determined by the following formula.

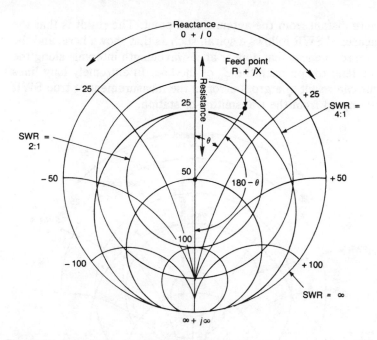

Fig. 11-15. Determining line lengths to obtain pure resistance at transmitter end of a feed line when the antenna contains reactance.

Physical Length of Feed Line

$$L = dv/(0.366f)$$

where L is the length in feet, v is the velocity factor, and f is the frequency in megahertz. For L in meters,

$$L = dv/(1.20f)$$

Once the reactance has been eliminated by this means, the resulting pure resistance can be matched to the transmitter output via a transformer.

In the actual case, the SWR does not change exactly according to the SWR circle as the measuring point is moved farther and farther from the antenna feed point. This is because any feed line contains some loss, and the reflected electromagnetic field is partly dissipated on the way back to the transmitter. Thus the SWR appears to decrease as the measuring point is made to be more and more distant from the antenna feed point. The result is that the measured SWR follows a spiral, such as that shown here, and the pure resistances that appear at ½-wavelength intervals along the line tend to approach the Z_0 of the line. In extremely long lines this can result in a gross error in the measurement of true SWR attempted from the transmitter or station.

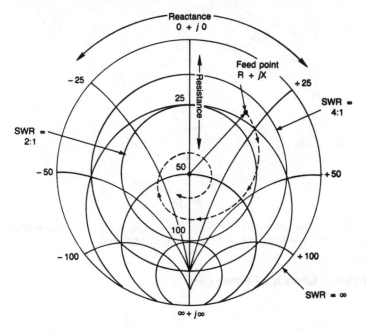

Fig. 11-16. Hypothetical example of SWR with distance from feed point in a line containing loss (a practical line).

Chapter 12

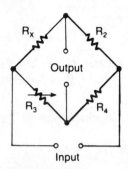

Miscellaneous Data

In this chapter you will find a helpful assortment of circuit and conversion formulas that do not fit conveniently into the preceding chapters.

Admittance in Terms of Impedance

$$Y = 1/Z$$

where Y = admittance (mhos)
Z = impedance (ohms)

Capacitive Susceptance

$$B_C = 1/X_C$$

where B_C = susceptance (mhos)
X_C = capacitive reactance (ohms)

Inductive Susceptance

$$B_L = \frac{X_L}{(R_L^2 + X_L^2)}$$

where B_L = susceptance (mhos)
   R_L = resistance of inductor (ohms)
   X_L = inductive reactance (ohms)

Conductance in Terms of Resistance

$$G = 1/R$$

where G = conductance (mhos)
   R = resistance (ohms)

Conductance in Terms of Current and Voltage

$$G = I/E$$
$$I = GE$$
$$E = I/G$$

where G = conductance (mhos)
   E = voltage
   I = current

Figure of Merit (Q) of Capacitor

$$Q = X_C/R = 1/(2\pi fCR)$$

where Q = figure of merit
   X_C = capacitive reactance
   C = capacitance
   f = frequency
   R = effective series resistance of capacitor

Figure of Merit (Q) of Inductor

$$Q = X_L/R = (2\pi fl)/R$$

where Q = figure of merit
   X_L = inductive reactance
   L = inductance
   f = frequency
   R = resistance of inductor

Figure of Merit (Q) in Terms of Resistance and Impedance

$$Q = \frac{\sqrt{Z^2 - R^2}}{R}$$

where Q = figure of merit
R = resistance
Z = impedance

Figure of Merit (Q)
in Terms of Test-Circuit Tuning Capacitances

$$Q = (2C_r)/(C_1 - C_2)$$

where Q = figure of merit
C_r = capacitance at resonance
C_1 = below-resonance capacitance at which circuit voltage is 0.707 of that at resonance
C_2 = above-resonance capacitance at which circuit voltage is 0.707 of that at resonance

Figure of Merit (Q) in Terms of Test-Circuit Frequencies

$$Q = f_r/(f_1 - f_2)$$

where f_r = resonant frequency
f_1 = below-resonance frequency at which circuit voltage is 0.707 of that at resonance
f_2 = above-resonance frequency at which circuit voltage is 0.707 of that of resonance

Quartz Crystal Frequency

$$f = K/t$$

where f = oscillation frequency (MHz)
t = thickness of plate (mils)
K = constant depending upon crystal cut (66.2 for AT cut, 100.78 for BT cut, 112.6 for X cut, and 77 for Y cut)

Frequency of Neon-Bulb Relaxation Oscillator

$$f = \frac{1}{RC \, ln\left[1 + \dfrac{V - V_m}{V - V_f}\right]}$$

where f = oscillation frequency (Hz)
ln = natural logarithm
V = dc supply voltage (volts)
V_f = firing voltage (volts)
V_m = maintaining voltage (volts)

Multivibrator Frequency

For symmetrical grid-plate-coupled circuit:

$$f = \frac{1000}{R_{g1}C_1 + R_{g2}C_2}$$

where f = frequency (kHz)
R_{g1} = R_{g2} = grid resistors (ohms)
C_1 = C_2 = grid coupling capacitors (μF)

For common-emitter transistor circuit, substitute R_b (base resistor) for R_g. For common-source FET circuit, substitute R_G (gate resistor) for R_g.

Voltage Amplification with Feedback

$$A_v = A(1 - A\beta)$$

where A_v = voltage amplification with feedback
A = voltage amplification without feedback
β = feedback factor (fraction of the amplifier output voltage fed back to the input)

Total Harmonic Distortion

$$D = \frac{\sqrt{h_2^2 + h_3^2 + h_4^2 + ... + h_n^2}}{f} \times 100$$

where D = total distortion (%)
f = amplitude of fundamental

h_2, h_3 ... h_n are amplitudes of indicated harmonics

Amplitude Modulation Percentage

$$M = \frac{E_m - E_c}{E_c} \times 100$$

where M = modulation percentage (%)
E_c = peak voltage of unmodulated carrier
E_m = peak voltage of modulated carrier

Electron-Beam Velocity

$$v = 3 \times 10^{10}\sqrt{1 - \left[\frac{1}{(2 \times 10^{-6})E + 1}\right]^2}$$

where v = velocity (cm/sec)
E = second anode accelerating voltage of CRT (volts)

Photon Energy

$$E = hf$$

where E = photon energy (J)
h = Planck's constant (6.6×10^{-34} J-sec)
f = frequency (Hz)

Temperature Conversion

Fahrenheit to Celsius: °C = (5/9) (°F − 32)
Celsius to Fahrenheit: °F = (9/5 °C) + 32
Kelvin to Celsius: °C = °K − 273.16
Celsius to Kelvin: °K = °C + 273.16

Gauge Factor for Strain Gauge

$$GF = (L\Delta R)/(R\Delta L)$$

where L = length
  R = resistance
  ΔL = change in length
  ΔR = change in resistance

Piezoelectric Constant

$$d_{ij} = g/F$$

where d_{ij} = piezoelectric constant
  g = generated charge
  F = applied force

Voltage Sensitivity of Piezoelectric Element

$$E = d_{ij}/d_c$$

where E = open-circuit voltage
  d_{ij} = piezoelectric constant
  d_c = dielectric constant

Charge Sensitivity of Transducer with Parallel Capacitance

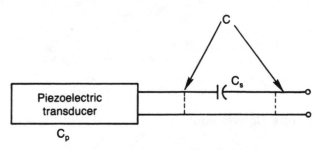

Fig. 12.1. Charge Sensitivity for piezoelectric transducer.

$$Q^* = (QC_s)/(C_p + C_s + C)$$

where Q^* = charge sensitivity with capacitance
Q = basic transducer charge sensitivity without capacitance
C_s = series capacitance
C_p = internal parallel capacitance of transducer
C = external parallel capacitance

Frequency Response of Transducer

$$k = fRF$$

where f = load frequency in hertz
R = input impedance of transducer in ohms
C = total capacitance, internal and external, in farads
when k < 1.0, the transducer response falls off rapidly. The relative response is fairly flat at values of k greater than 1.0.

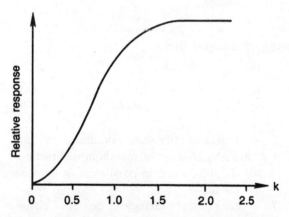

Fig. 12-2. Relative response for piezoelectric transducer as a function of value of k.

Cross-Axis Sensitivity of Piezoelectric Transducer

$$Q_t/Q_{xx} = 100 \tan \theta \cos \phi$$

where Q_t/Q_{xx} = cross-axis sensitivity as a percentage of input-axis sensitivity
θ = angle of measurement relative to input axis
ϕ = angle of load axis relative to polarized axis

163

Energy Emitted by Black Body

$$\lambda = 2898/T$$

where λ = wavelength in microns where emitted energy is
strongest

T = temperature in degrees Kelvin

Thermocouple Voltage

$$\log V = A \log (T_1 - T_2) + B(T_1^2 - T_2^2)$$

where V = generated voltage in microvolts
T_1 = hot-junction temperature
T_2 = cold-junction temperature
A,B are constants that depend on metals used in the
device

Deflection of Bimetal Strip

$$R = \frac{(t_1 + t_2)^3}{6dTt_1t_2}$$

where R = radius of curvature in millimeters
t_1,t_2 are thicknesses of metals in millimeters
d = the difference in coefficient of expansion of two
metals
T = temperature change in degrees Celsius or Kelvin

Also,

$$D = KTL^2/t$$

where D = actual deflection in millimeters
t = total thickness of strip $(t_1 + t_2)$ in millimeters
K = flexivity × 0.5
T = temperature change in degrees Celsius or Kelvin
L = length of strip in millimeters

Force Exerted by Bimetal Strip

$$F = 4ETwt^2/L$$

where F = force in ounces
E = elastic modulus of strip in pounds per square inch
w, t, L are width, thickness and length of strip in inches

Torque for Coiled Bimetal Strip

$$F = 4700Twt^2$$

where T = temperature difference in degrees Fahrenheit
w = strip width in inches
t = strip thickness in inches

Deflection of Spring

$$S = 8FD^3N/Gd^4$$

where F = force in newtons
D = mean diameter of spring, meters
d = diameter of spring wire, meters
G = modulus of rigidity of spring material
N = number of coils in spring
S = deflection, meters

Torsional Stress of Spring

$$T = 8FD/\pi d^3$$

where T = torsional stress
F = force in newtons
D = mean diameter of spring, meters
d = diameter of spring wire, meters

Torque vs. Electric Current in Coil

$$Q = ABnI/9.810$$

where Q = torque, gram-centimeters
$\quad\quad\quad A$ = area of coil in square centimeters
$\quad\quad\quad B$ = flux density in lines per square centimeter
$\quad\quad\quad n$ = number of turns in coil
$\quad\quad\quad I$ = current in milliamperes

Sound Pressure

$$P = 2.9 \times 10^{-9} \times 10^{dB/20}$$

where $\quad P$ = pressure in pounds per square inch
$\quad\quad\quad$ dB = sound intensity in decibels relative to threshold of hearing

Absorptance

$$A = 10 \log (P_1/P_2)$$

where $\quad A$ = absorptance, dB
$\quad\quad\quad P_1$ = impinging power, watts
$\quad\quad\quad P_2$ = output power, watts

Accuracy

$$A = 100 \mid x_s - x_t \mid /x_s$$

where $\quad A$ = accuracy, percent
$\quad\quad\quad x_s$ = reading of standard instrument
$\quad\quad\quad x_t$ = reading of instrument under test, same units as x_s

Albedo

$$A = 100e_r/e_i$$

where $\quad A$ = albedo, percent
$\quad\quad\quad e_r$ = reflected energy
$\quad\quad\quad e_i$ = incident energy

Angular Frequency

$$f_{d/s} = 360f_{Hz}$$

$$f_{r/s} = 2\pi f_{Hz}$$

where $f_{d/s}$ = frequency in degrees per second
$\quad\quad\;\; f_{r/s}$ = frequency in radians per second
$\quad\quad\;\; f_{Hz}$ = frequency in hertz

Audibility

$$A = 160 + 10 \log X$$

where A = audibility in decibels
$\quad\quad\;\; X$ = sound intensity in watts per square centimeter

Avalanche Impedance of PN Junction

$$Z_A = E_R / I_R$$

where Z_A = avalanche impedance, ohms
$\quad\quad\;\; E_R$ = reverse voltage
$\quad\quad\;\; I_R$ = reverse current

BCD Notation

Digit in Base 10	BCD Notation
0	0000
1	0001
2	0010
3	0011
4	0100
5	0101
6	0110
7	0111
8	1000
9	1001

Fig. 12-3. The digits 0 through 9 and their binary-coded decimal equivalents.

Boltzmann Constant

The *Boltzmann constant* is a coefficient that defines the relationship between temperature and electron energy. The relationship is linear and is determined by this constant, usually abbreviated by the letter k. The higher the temperature, the greater the amount of energy in an electron.

The most common expressions are

$$k = 1.38 \times 10^{-28} \text{ J/K}$$
$$= 8.61 \times 10^{-5} \text{ } e\text{V/K}$$

where J is the abbreviation for joules, eV is the abbreviation for electron volts and K represents degrees Kelvin.

Conductivity of Copper Wire

American Wire Gauge	MilliSiemens per Meter
2	1,913,375
4	1,203,319
6	756,705
8	475,879
10	299,411
12	188,265
14	118,369
16	74,451
18	46,820
20	29,449
22	18,518
24	11,647
26	7,323
28	4,606
30	2,897
32	1,822
34	1,146

Fig. 12-4. Conductivity in milliSiemens per meter of various gauges of copper wire.

Cosine Law

The energy intensity from a perfectly diffusing surface is greatest in the direction perpendicular to that surface. The intensity varies with the cosine of the angle θ relative to the normal. The intensity also varies with the sine of the angle ϕ relative to the surface, where $\theta = 90° - \phi$.
Thus

$$P = \cos \theta = \cos (90° - \phi) = \sin \phi = \sin (90° - \theta)$$

where P is the intensity on a relative scale, with the maximum possible intensity being 1.

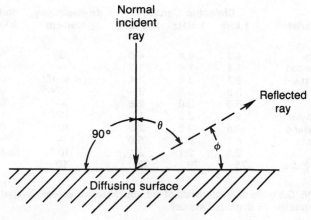

Fig. 12-5. Cosine law.

Damping Factor of Coil

$$a = R/(2L)$$

where a = damping factor
R = radio-frequency resistance of coil, ohms
L = coil inductance, henrys

DeBroglie Equation

$$\lambda = (6.54 \times 10^{-27})/(mv)$$

where λ = wavelength, centimeters
m = rest mass of object in grams
v = velocity of object in centimeters per second

This equation holds for nonrelativistic speeds, or velocities of less than about 0.1 times the speed of light. At relativistic speeds, the wavelength is multiplied by the factor $\sqrt{1 - v^2/c^2}$, where c is the speed of light.

169

Dielectric Constants

Material	Dielectric Constant 1 kHz	1 MHz	100 MHz	dc Resistivity, ohm-cm	Rating, kV/mm
Bakelite	4.7	4.4	4.0	10^{11}	0.1
Balsa wood	1.4	1.4	1.3	—	—
Epoxy resin	3.7	3.6	3.4	4×10^7	0.13
Fused quartz	3.8	3.8	3.8	10^{19}	0.1
Paper	3.3	3.0	2.8	—	0.07
Polyethylene	2.3	2.3	2.3	10^{17}	1.4
Polystyrene	2.6	2.6	2.6	10^{18}	0.2
Porcelain	5.4	5.1	5.0	—	—
Teflon	2.1	2.1	2.1	10^{17}	6
Water (pure)	78	78	78	10^6	—

Fig. 12-6. Dielectric characteristics of various materials at room temperature (approximately 25 degrees Celsius).

Dissipation Factor of Dielectric Material

$$D = \tan \theta = \tan (90° - \phi)$$

where D = dissipation factor
θ = loss angle
ϕ = phase angle

Droop in a Pulse

$$D = 100 (E_1 - E_2)/E_1$$

where D = droop
E_1 = maximum pulse amplitude
E_2 = amplitude just before rapid drop begins

Duty Cycle

$$D = 100t/t_0$$

where D = duty cycle, percent

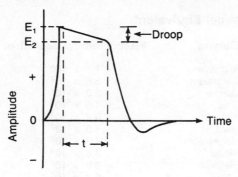

Fig. 12-7. Droop.

$$t = \text{total length of time circuit is in operation during time } t_0$$
$$t_0 = \text{measurement time}$$

Duty Factor

$$DF = 100P_{av}/P_{max}$$

where DF = duty factor, percent
P_{av} = average power
P_{max} = peak power

In a pulse train,

$$DF = 100tf$$

where t = duration time of pulse train
f = pulse frequency in hertz

Effective Radiated Power

$$ERP = P \times 10^{(A_p/10)}$$

where ERP = effective radiated power in watts
P = power reaching antenna in watts
A_p = antenna power gain in decibels

171

Electrochemical Equivalent

Element	Electrochemical Equivalent
Aluminum	1.1×10^4
Chromium	5.6×10^3
Copper	3.0×10^3
Gold	1.5×10^3
Iron	5.2×10^3
Lead	1.9×10^3
Magnesium	7.9×10^3
Nickel	3.3×10^3
Platinum	2.0×10^3
Silver	8.9×10^2
Tin	3.3×10^3
Zinc	3.0×10^3

Fig. 12-8. Electrochemical equivalents in coulombs per gram for some common metals.

Electromagnetic Spectrum

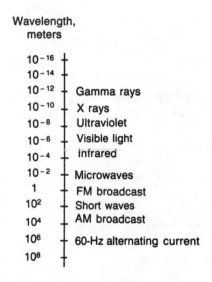

Fig. 12-9. Electromagnetic spectrum (logarithmic scale).

Exponential Distribution

The exponential distribution is a form of statistical distribution used for determining the probability that an event will occur

172

within a specified time interval. The probability distribution is given in general by

$$P = fe^{-ft}$$

where P = probability
f = frequency of occurrence
t = time interval
e is a constant, approximately 2.718

Feedback Ratio

$$F = e_f/e_o$$

where e_f = feedback voltage
e_o = output voltage

Free-Space Propagation Loss

$$F = k/d$$

where F = field strength, volts per meter
k = constant depending on source intensity
d = distance from source, meters

Hall Constant

In a current-carrying conductor, the transverse electric field, the magnetic field, and the current density are mutually dependent according to the following relation:

$$e/(im) = k$$

where e = intensity of transverse electric field
i = current density
m = magnetic field strength
k = constant, called the Hall constant

Horizon Distance

$$d = \sqrt{1.53h}$$

where d = visual horizon distance
h = height above flat terrain

For vhf and uhf radio waves, the formula is approximately

$$d = \sqrt{2h}$$

For a complete radio circuit,

$$d = \sqrt{2g} + \sqrt{2h}$$

where g and h are the heights of the radio antennas at the respective stations

Interferometer Resolution

$$a = 57.3\lambda/L$$

where a = angular separation between lobes in degrees
λ = wavelength in meters
L = spacing in meters between antennas

Modulation Coefficient

$$m = (E_m - E_c) / E_c$$

where m = modulation coefficient
E_m = maximum peak-to-peak voltage of modulated carrier
E_c = peak-to-peak voltage of unmodulated carrier

This may also be expressed in percent as

$$m = 100(E_m - E_c) / E_c$$

Modulation Index

$$m = f/g$$

where m = modulation index

174

f = maximum instantaneous frequency deviation, kHz

g = instantaneous audio modulating frequency, kHz

This may be expressed as a percentage by

$$m = 100f/g$$

Modulation Percentage

The modulation percentage is defined as the percent form of the modulation coefficient.

Monostable Multivibrator Pulse Duration

The pulse duration is given by

$$T = 0.69RC$$

where T is the duration in microseconds

R resistor value in ohms

C capacitor value in microfarads

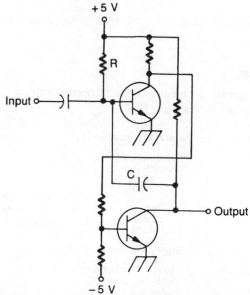

Fig. 12-10. Monostable multivibrator pulse time.

175

Noise Figure of Amplifier or Receiver

Suppose that the noise output of an ideal network results in a signal-to-noise ratio $R_1 = P/Q_1$, where P is the signal power and Q_1 is the noise power. Let the noise power generated in the actual circuit be Q_2, so that the actual signal-to-noise ratio is $R_2 = P/Q_2$. Then the noise figure in decibels, N, is given by

$$N = 10 \log (R_1/R_2) = 10 \log (Q_2/Q_1)$$

A perfect circuit has a noise figure of 0 dB. In practice this value is never achieved since every circuit generates some internal noise.

Peak Power

In a sine-wave ac signal, the peak power is determined from the rms voltage, E_{rms}, and the rms current, I_{rms}:

$$P_{pk} = 2E_{rms}I_{rms}$$

where P_{pk} represents the peak power. In terms of peak voltage and current, E_{pk} and I_{pk},

$$P_{pk} = E_{pk}I_{pk}$$

Voltage is expressed in volts, current in amperes, and power in watts.

Permeability

The magnetic permeability for various common materials is shown in the accompanying table.

Substance	Permeability (Approx.)
Aluminum	Slightly more than 1
Bismuth	Slightly less than 1
Cobalt	60-70
Ferrite	100-3000
Free Space	1
Iron	60-100
Iron, refined	3000-8000
Nickel	50-60
Permalloy	300-30,000
Silver	Slightly less than 1
Steel	300-600
Super-permalloys	100,000-1,000,000
Wax	Slightly less than 1
Wood, dry	Slightly less than 1

Fig. 12-11. Permeability factors of some common substances.

Photon Energy

$$e = hf$$

where e = energy in ergs
h = Planck's constant = 6.62×10^{-27} erg seconds
f = frequency in hertz

If the wavelength in meters is given by λ, then

$$e + 3 \times 10^8 \times h/\lambda$$

Poisson Distribution

The Poisson distribution is a form of normal distribution used for determining the average number of random events that take place within a given amount of time. During a specified time interval t, in seconds, the probability that an event will take place n times is given by

$$P = \left[(ft)^n/n!\right]e^{-ft}$$

where f is the average frequency of the event (events per second) and P is the probability. The value of e is about 2.718.

Rayleigh Distribution

The Rayleigh distribution is a form of normal distribution in two dimensions. Probability density is determined according to the distance of a point (x,y) from the origin of the Cartesian plane. In polar coordinates the probability density for a point (r,θ) depends on the value of r. The Rayleigh function ϕ is given by

$$\phi(r) = (r/s^2) \, e^{-\left[(r^2/2)s^2\right]}$$

where s is the standard deviation.

Regulation in Power Supply

$$R = 100(E_2 - E_1) / E$$

where E = optimum voltage (rated voltage)

$$E_1 = \text{minimum voltage}$$
$$E_2 = \text{maximum voltage}$$

Plus-or-minus regulation is given by

$$R_\pm = \pm\ 100(E_2 - E)/E \text{ or}$$
$$R_\pm = \pm\ 100(E - E_1)/E$$

where R and $R_\pm$ are the regulation values in percent.

Resistance of Solid Copper Wire

AWG	Ohms/km	AWG	Ohms/km
1	0.42	21	43
2	0.52	22	54
3	0.66	23	68
4	0.83	24	86
5	1.0	25	110
6	1.3	26	140
7	1.7	27	170
8	2.1	28	220
9	2.7	29	270
10	3.3	30	350
11	4.2	31	440
12	5.3	32	550
13	6.7	33	690
14	8.4	34	870
15	11	35	1100
16	13	36	1400
17	17	37	1700
18	21	38	2200
19	27	39	2800
20	34	40	3500

Fig. 12-12. Resistance of solid copper wire for American Wire Gauge (AWG) 1 through 40 in ohms per kilometer.

Resistance Loss vs. Efficiency

$$E = 100S/(R + S)$$

where E = efficiency in percent
R = loss resistance
S = load resistance

Resistor Color Code

(A) Color	Digit
Black	0
Brown	1
Red	2
Orange	3
Yellow	4
Green	5
Blue	6
Violet	7
Gray	8
White	9

(B) Color	Multiplier
Black	1
Brown	10
Red	10^2
Orange	10^3
Yellow	10^4
Green	10^5
Blue	10^6
Gold	0.1
Silver	0.01

Fig. 12-13. Resistor color code. (A) First and second significant figure bands; (B) multiplier band; (C) tolerance (%); (D) failure rate.

(C) Tolerance (%)	Color
Gold ±5	Brown
Silver ±10	Red
None ±20	Orange
	Yellow

(D) Failure Rate
%/1000 hours

1
0.1
0.01
0.001

Sag of Wire Span or Cable Span

$$s = 0.612 \sqrt{mn - m^2}$$

where s = sag

m = distance between supports

n = length of cable (all measurements in same units)

Selectance

$$S = (E_2 - E_1) / (f_2 - f_1) \text{ if } f_2 > f_1$$
$$S = (E_2 - E_1) / (f_1 - f_2) \text{ if } f_2 < f_1$$

where S = selectance
E_1 = voltage needed for 10-dB signal-to-noise ratio at frequency f_1
E_2 = voltage needed for 10-dB signal-to-noise ratio at frequency f_2

Signal-plus-noise-to-noise Ratio

$$(S + N)/N = 20 \log \left[(E_s + E_n)/E_n \right]$$

where E_s = signal voltage
E_n = noise voltage

Signal-to-noise Ratio

$$S/N = 20 \log (E_s/E_n)$$

where E_s = signal voltage
E_n = noise voltage

Snell's Law

$$\sin \theta / \sin \phi = c$$

where θ = angle relative to normal in air
ϕ = angle relative to normal in refractive substance

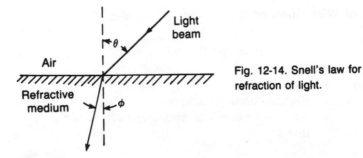

Fig. 12-14. Snell's law for refraction of light.

Thermal Conductivity

Element	Thermal conductivity, Milliwatts per meter per degree Celsius
Aluminum	22
Carbon	2.4
Chrominum	6.9
Copper	39
Gold	30
Iron	7.9
Lead	3.5
Magnesium	16
Mercury	0.85
Nickel	8.9
Platinum	6.9
Silicon	8.4
Silver	41
Thorium	4.1
Tin	6.4
Tungsten	20
Zinc	11

Fig. 12-15. Thermal conductivity of some common elements.

Transformer Efficiency

$$Eff = 100E_sI_s/(E_pI_p)$$

where Eff = efficiency, percent
E_s = secondary voltage
I_s = secondary current
E_p = primary voltage
I_p = primary current

Trapezoidal Pattern
Determination of Modulation Percentage

$$m = 100(L - S)/(L + S)$$

where m = modulation percentage (AM)
L = length of long side of trapezoid
S = length of short size of trapezoid

This applies for modulation percentages between zero and 100, but not greater than 100.

Volume Units

$$VU = 10 \log (P/2.51)$$

where VU = level in volume units
P = audio power in milliwatts

Chapter 13

English
Letter Symbols

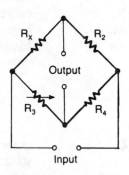

A

A	amperes; area
a	prefix *atto-* (10^{-18})
aA	attoamperes
AC	alternating current
AF	audio frequency
AFC	automatic frequency control
Ah	ampere-hours
aH	attohenrys
A_i	current amplification
AM	amplitude modulation; amplitude modulated
ALC	automatic level control
A_p	power amplification
aV	attovolts
A_v	voltage amplification
AVC	automatic volume control (also AGC)
aW	attowatts

B

β	beta; h_{FE}; H_{fe}
B	base; battery; bels; magnetic flux density; complementary MOS

b	base
B_c	capacitive susceptance
B_i	input susceptance
B_L	inductive susceptance
B_o	output susceptance
BV	breakdown voltage
BW	bandwidth

C

C	capacitance; capacitor; Celsius; collector; constant; coulombs
c	collector; speed of light (300,000 km/sec or 186,000 mi/sec); prefix *centi-* (10^{-2})
C_c	collector capacitance
C_D	drain capacitance
C_G	gate capacitance
C_g	grid capacitance
C_{gk}	grid-cathode capacitance
C_{gp}	grid-plate capacitance
CCW	counterclockwise
C_i	input capacitance
C_J	junction capacitance
CML	current-mode logic
CMOS	complementary MOS
CMRR	common-mode rejection ratio
C_n	neutralizing capacitance; null capacitance; remote value of capacitance
C_o	output capacitance
cos	cosine of angle
cosech	hyperbolic cosecant
cosh	hyperbolic cosine
cot	cotangent of angle
coth	hyperbolic cotangent
C_p	plate capacitance
C_{pk}	plate—cathode capacitance
CRT	cathode-ray tube
C_s	source capacitance
C_s	screen capacitance
csc	cosecant of angle

184

C_t	total capacitance
CW	continuous wave; clockwise

D

D	diameter; diffusion constant; diode; displacement; dissipation factor; distance; drain; electrostatic flux density
d	diameter; differential; distance, thickness; prefix *deci-* (0.1)
da	prefix *deka-* (10)
dB	decibels
dBm	decibels referred to 1 mW
dBf	decibels referred to 1 femtowatt
DC	direct current
dcc	double cotton-covered (wire)
DCWV	DC working voltage
DCTL	direct-coupled transistor logic
DMM	digital multimeter
DPDT	double-pole double-throw (switch)
DPST	double-pole single-throw (switch)
DSB	double sideband
dsc	double silk-covered (wire)
DTL	diode/transistor logic
DVM	digital voltmeter

E

E	electric field intensity, emitter, voltage
e	electron charge (1.6×10^{-19} coulomb); natural number (2.71828...); voltage
E_a	anode voltage
E_{ac}; E_{AC}	AC voltage
E_{avg}	average voltage
E_{bb}	plate supply voltage
E_c	grid-bias voltage
ECL	emitter-coupled logic
E_{dc}; E_{DC}	DC voltage
E_f	filament voltage; fundamental-frequency voltage

E_g	grid voltage; generator voltage
E_h	heater voltage
E_i	input voltage
E_{inst}	instantaneous voltage
E_k	cathode voltage
E_L	load voltage; voltage across inductor
E_m	maximum voltage
E_{max}	maximum voltage
E_{min}	minimum voltage
E_n	neutralizing voltage; null voltage
E_o	output voltage; reference voltage
E_p	plate voltage; peak voltage
E_{p-p}	peak-to-peak voltage
E_r	reverse voltage
E_{rms}	root-mean-square (effective) voltage
E_s	screen voltage
E_{sup}	suppressor voltage
EVM	electronic voltmeter
exp	exponential; exponent

F

F	electrostatic field intensity; farads; force
f	frequency; force; fundamental, prefix *femto-* (10^{-15})
f_α	alpha cutoff frequency
fA	femtoamperes
f_c	center frequency; cutoff frequency
f_{co}	cutoff frequency
FET	field-effect transistor
fH	femtohenrys
f_i	input frequency
FM	frequency modulation; frequency modulated
f_m	modulation frequency
f_{max}	maximum frequency
f_{min}	minimum frequency
f_o	output frequency; reference frequency
f_r	resonant frequency
f_t	Unity-gain cutoff frequency; gain-bandwidth product
fV	femtovolts
fW	femtowatts

G

G	conductance; gate; grid; prefix *giga-* (10^9)
g	acceleration due to gravity; grid
Ge	germanium
g_{fs}	common-source forward transconductance
GHz	gigahertz
G_i	input conductance
g_m; G_m	transconductance
GND	ground
G_o	output conductance
g_{oss}	common-source output conductance
grad	gradient of
GV	gigavolts
GW	gigawatts

H

H	henrys; magnetic field intensity
h	harmonic; height; Planck's constant (6.547×10^{-27} erg-sec); prefix *hecto-* (100)
HF	high frequency
h_{fb}	common-base forward-current transfer ratio
h_{fc}	common-collector input resistance
h_{FE}	common-emitter forward current transfer ratio (DC beta)
H_{fe}	AC beta
h_{ib}	common-base input resistance
h_{ic}	common-collector forward-current transfer ratio
h_{ie}	common-emitter input impedance
HNIL	high-noise-immunity logic
h_{ob}	common-base output conductance
h_{oc}	common-collector output conductance
h_{oe}	common-emitter output admittance
h_{rb}	common-base reverse-voltage transfer ratio
h_{rc}	common-collector reverse-voltage transfer ratio
h_{re}	common-emitter reverse-voltage transfer ratio
HTL	high-threshold logic
Hz	hertz (cycles per second)
h_{11}	input impedance
h_{12}	reverse-voltage transfer ratio

h_{21}	forward-current transfer ratio
h_{22}	output admittance

I

I, i	current
I_{ac}	alternating current
I_{af}	audio-frequency current
I_{avg}	average current
I_b, I_B	base current
IC	integrated circuit
$I_c; I_C$	capacitor current; collector current
I_{co}	cutoff current
ICW	interrupted continuous wave
I_D	drain current
I_{dc}	direct current
IDC	instantaneous deviation control (Motorola)
$I_e; I_E$	emitter current
I_F	forward current
I_f	feedback current; filament current; fundamental-frequency current
IF	intermediate frequency
I_G	gate current
I_g	grid current
IGFET	insulted-gate field-effect transistor (MOSFET)
I_h	heater current
I_i	input current
IIL, I^2L	integrated injection logic
I_{inst}	instantaneous current
I_k	cathode current
I_L	load current; inductor current
I_m	maximum current
I_{max}	maximum current
I_{min}	minmum current
I_o	output current; reference current
I_p	plate current; peak current
ips	inches per second
I_r, I_R	reverse current; resistor current
I_{rf}	radio-frequency current
I_{rms}	root-mean-square (effective) current
I_S	source current
I_s	current source; screen current

I_{sup}	suppressor current
I_t	total current
I_x	current through reactance
I_z	zener current; current through impedance

J

J	jack; joules; Poynting's vector
j	complex operator ($\sqrt{-1}$)
JFET	junction field-effect transistor

K

K	Kelvin; cathode; relay
k	cathode; a constant; prefix *kilo-* (1000)
kA	kiloamperes
kHz	kilohertz
kV	kilovolts
kVA	kilovoltamperes
kVAR	reactive kilovoltamperes
kW	kilowatts
kWh	kilowatt-hours

L

L	inductance; inductor
l	inductance; length
LC	inductance—capacitance
LCR	inductance—capacitance—resistance
LF	low frequency
ln; $\log_e$	natural logarithm
log; $\log_{10}$	common logarithm
L_p	primary inductance
LR	inductance—resistance
L_s	secondary inductance
LSB	lower sideband; least significant bit

M

M	modulation factor; mutual inductance; prefix *mega-* or *meg* $^{(10^6)}$
m	mass; meters; prefix *milli-* (10^{-3})

MA	megamperes
mA	milliamperes
MADT	microalloy diffused transistor
MAT	microalloy transistor
MCW	modulated continuous wave
mfd	microfarad
mH	millihenrys
MHz	megahertz
mm	millimeters
mm Hg	See *torr*
MMF	magnetomotive force
MOS	metal oxide semiconductor
MOSFET	metal-oxide-semiconductor field-effect transistor
MOST	metal-oxide-semiconductor transistor
ms; msec	milliseconds
MSB	most significant bit
MTL	merged-transistor logic
MV	megavolts
mV	millivolts
MW	megawatts
mW	milliwatts

N

N	number; negative N-type semiconductor material
n	number; prefix *nano-* (10^{-9})
nA	nanoamperes
NC	no connection, normally closed
N_D	noise density
NF	noise figure
nF	nanofarads
NFET	N-channel field-effect transistor
nH	nanohenrys
NMOS	N-channel MOS transistor
No.	number
NO	normally open
N_p	number of primary turns
NPN	negative-emitter—positive-base—negative collector bipolar transistor
NPNP	four-layer semiconductor device
NP0; NPO	zero temperature coefficient

N_s	number of secondary turns
ns	nanoseconds
NTC	negative temperature coefficient
nV	nanovolts
nW	nanowatts

P

P	power; positive; P-type semiconductor material
p	sound pressure; prefix *pico-* (10^{-12})
PA	power amplifier; public address system
pA	picoamperes
P_{avg}	average power
PC	photocell; point contact; printed circuit
PCM	pulse code modulation
P_c	collector power
P_D	power dissipation
P_D	drain power
PDM	pulse data modulation
P_{DM}	maximum permissible power dissipation
PF	power factor
pF	picofarads
P_f	feedback power; filament power
PFET	P-channel field-effect transistor
P_G	generator power
P_g	grid power
pH	picohenrys
P_h	heater power
pi	constant π (3.14159...)
P_i	input power
PIN	Positive—instrinsic—negative semiconductor
P_L	load power
PM	phase modulation
PMOS	P-channel MOS transistor
P_{max}	maximum power
P_{min}	minimum power
P_n	noise power
PNP	positive-emitter—negative-base—positive-collector bipolar transistor
PNPN	four-layer semiconductor device
P_o	output power; reference power
P_p	plate power

ppm	parts per million
PPM	pulse-position modulation
P_S	source power
P_s	screen power
ps; psec	picoseconds
PTC	positive temperature coefficient
pV	picovolts
pW	picowatts
P_Z	zener power

Q

Q	electric charge; quantity of electricity; selectivity; figure of merit; transistor
q	electric charge

R

R	resistance; resistor
$\mathcal{R}$	reluctance
r	radius
R_b	base resistance
RC	resistance—capacitance
R_c	collector resistance
R_D	drain resistance
R_e	emitter resistance
rev	revolutions
R_F	forward resistance
RF	radio frequency
R_f	feedback resistance; filament resistance
RFC	radio-frequency choke
R_G	gate resistance; generator resistance
R_g	grid resistance
R_h	heater resistance
R_i	input resistance
R_k	cathode resistance
RL	resistance—inductance
R_L	load resistance
RLC	resistance—inductance—capacitance
r_m	mutual transfer characteristic of bipolar transistor
RMS	root-mean-square (effective) value

R_p	parallel resistance; plate resistance; primary resistance
rpm	revolutions per minute
rps	revolutions per second
R_R	reverse resistance
R_r	radiation resistance of antenna
R_{rf}	radio-frequency resistance
R_S	source resistance
R_s	screen resistance; secondary resistance; series resistance
R_{sup}	suppressor resistance
R_t	total resistance
RTL	resistor/transistor logic
R_Z	zener resistance
R_{11}	input resistance
R_{12}	reverse transfer resistance
R_{21}	forward transfer resistance
R_{22}	output resistance

S

S	sensitivity; source; switch
SBDT	surface-barrier diffused transistor
SBT	surface-barrier transistor
scc	single cotton-covered (wire)
sce	single cotton enameled (wire)
SCR	silicon controlled rectifier
SCS	silicon controlled switch
sec	secant of angle
sech	hyberbolic secant
SHF	superhigh frequency
Si	silicon
sin	sine of angle
sinh	hyeprbolic sine
SPDT	single-pole, double-throw (switch)
SPST	single-pole, single-throw (switch)
SSB	single sideband
ssc	single silk-covered (wire)
SW	shortwave; switch

T

T	temperature; thermistor; transformer; torr (mm

	Hg); prefix *tera-* (10^{12}); total
t	time
τ	time constant
tan	tangent of angle
tanh	hyperbolic tangent
TC	temperature coefficient; thermocouple
T_d	delay time
TTL	transistor/transistor logic
TV	television
TVM	transistorized voltmeter
TW	terawatts

U

u	prefix *micro-*; microns; substitute for Greek letter *mu*; transconductance
UHF	ultrahigh frequency
UJT	unijunction transistor
ULF	ultralow frequency
USB	upper sideband

V

V	tube; varistor; voltage; volts
VA	voltamperes
VAr	voltamperes reactive
V_{be}	base—emitter voltage
V_b	base voltage
V_{BB}	base supply voltage
V_c	collector voltage
V_{cb}	collector—base voltage
V_{CC}	collector supply voltage
V_{ce}	collector—emitter voltage
VCA	voltage-controlled amplifier
VCO	voltage-controlled oscillator
VCX	voltage-controlled crystal oscillator
V_D	drain voltage
V_{DD}	drain supply voltage
VDR	voltage-dependent resistor
V_e	emitter voltage
V_{eb}	emitter—base voltage
V_{EE}	emitter supply voltage

194

V_F	forward voltage
V_G	gate voltage; generator voltage
VHF	very high frequency
VLF	very low frequency
VOM	voltohmmeter
V_n; V_N	noise voltage
vol	volume
V_P	gate—source pinchoff voltage
V_p	peak voltage
VR	voltage regulator
V_R	reverse voltage
V_S	source voltage
V_{SS}	source supply voltage
VTVM	vacuum-tube voltmeter
V_Z	zener voltage

W

W	watts; width; work function
w	weight; work
Wh	watt-hours
WL; λ	wavelength
W s; W sec	watt-seconds
wt	weight
WV	working voltage
WW	wirewound

X

X	reactance; crossover; crystal
x	reactance; horizontal axis of graphs and screens; unknown quantity
X_C	capacitive reactance
X_L	inductive reactance
X_t	total reactance
XTAL	crystal

Y

Y	admittance

y	admittance; vertical axis of graphs and screens
Y_i	input admittance
Y_o	output admittance
Y_S	source admittance

Z

Z	impedance
Z_b	base impedance
Z_c	collector impedance
Z_D	drain impedance
Z_e	emitter impedance
Z_F	forward impedance
Z_G	gate impedance; generator impedance
Z_i	input impedance
Z_o	output impedance
Z_p	parallel impedance; plate impedance
Z_S	source (generator) impedance; series impedance
Z_Z	zener impedance

Chapter 14

Greek
Letter Symbols

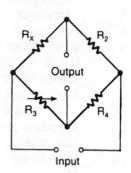

The Greek Alphabet

Greek Capital Letter	Greek Lowercase Letter	Greek Name	Greek Capital Letter	Greek Lowercase Letter	Greek Name
A	α	Alpha	N	ν	Nu
B	β	Beta	Ξ	ξ	Xi
Γ	γ	Gamma	O	o	Omicron
Δ	δ	Delta	Π	π	Pi
E	ϵ	Epsilon	P	ϱ	Rho
Z	ζ	Zeta	Σ	σ	Sigma
H	η	Eta	T	τ	Tau
Θ	θ	Theta	Υ	υ	Upsilon
I	ι	Iota	Φ	ϕ	Phi
K	$\varkappa$	Kappa	X	χ	Chi
Λ	λ	Lambda	Ψ	ψ	Psi
M	μ	Mu	Ω	ω	Omega

α Angle; attenuation factor; common-base current amplification factor; radiation

β Angle; common-emitter amplification factor

Γ Complex propagation constant (Hertzian vector)

γ Angle; Euler's constant; propagation constant

Δ Decrement; determinant; increment; permittivity

δ Angle; decrement; density; Dirac impulse function

ϵ	Inductivity; natural number (2.71828...); permittivity
η	Efficiency; hysteresis; surface charge density
Θ	Angle; temperature coefficient
θ	Angle; phase displacement; time constant
Λ	Permeance
λ	Attenuation constant; meter(s); wavelength
μ	amplification factor; permeability; prefix *micro*- (10^{-6})
μA	Microampere(s)
μ_e	Electron mobility
μF	Microfarad(s)
μH	Microhenry(s)
μ_h	Hole mobility
$\mu\Omega$	Microhm(s)
μsec	Microsecond(s)
μV	Microvolt(s)
μW	Microwatt(s)
ν	Frequency; reluctivity
Π	Hertzian vector; product
π	Constant 3.14159...; number of radians in 180°
ϱ	Resistivity; volume charge density
Σ	Summation
σ	Complex propagation constant; electrical conductivity; leakage coefficient; standard deviation; surface charge density
τ	Carrier lifetime; density; time constant; time—phase displacement; transmission factor
Φ	Magnetic flux; scalar potential
ϕ	Angle; contact potential; magnetic flux; phase
χ	Angle; susceptibility
ψ	Angle; dielectric flux; phase difference
Ω	Ohm(s)
$G\Omega$	Gigohm(s)
$k\Omega$	Kilohm(s)
$M\Omega$	Megohm(s)
$m\Omega$	Milliohm(s)
$T\Omega$	Teraohm(s)
ω	Angular velocity

Chapter 15

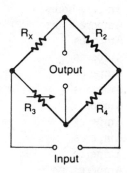

Conversion Factors

To Convert From	To	Multiply By
Abamperes	Amperes	10
Abamperes	Statamperes	2.9974×10^{10}
Abampere-turns	Ampere-turns	10
Abampere-turns	Gilberts	12.57
Abcoulombs	Coulombs	10
Abcoulombs	Faradays	10^{-4}
Abcoulombs	Statcoulombs	2.9974×10^{10}
Abhenrys	Henrys	1×10^{-9}
Abhenrys	Stathenrys	1.113×10^{-21}
Abmhos	Mhos	10^{9}
Abmhos	Micromhos	10^{15}
Abohms	Ohms	10^{-9}
Abohms	Statohms	1.113×10^{-21}
Abvolts	Statvolts	3.336×10^{-11}
Abvolts	Volts	10^{-8}
Acres	Ares	40.469
Acres	Hectares	0.4047
Acres	Sq centimeters	40,468,726
Acres	Sq feet	43,560
Acres	Sq inches	6.27264×10^{6}
Acres	Sq kilometers	4.047×10^{-3}
Acres	Sq meters	4046.9

To Convert From	To	Multiply By
Acres	Sq miles	1.56×10^{-3}
Acres	Sq yards	4840
Ampere-hours	Faradays	0.03731
Ampere-hours	Milliampere-hours	10^3
Amperes	Abamperes	0.1
Amperes	Attoamperes	10^{18}
Amperes	Coulombs/sec	1
Amperes	Femtoamperes	10^{15}
Amperes	Kiloamperes	10^{-3}
Amperes	Megamperes	10^{-6}
Amperes	Microamperes	10^6
Amperes	Milliamperes	10^3
Amperes	Nanoamperes	10^9
Amperes	Picoamperes	10^{12}
Amperes/volt	Microamperes/volt	10^6
Amperes/volt	Milliamperes/volt	10^3
Ampere-turns	Abampere-turns	0.1
Angstrom units	Feet	4.724×10^{-8}
Angstrom units	Centimeters	10^{-8}
Angstrom units	Inches	3.937×10^{-9}
Angstrom units	Meters	10^{-10}
Angstrom units	Microangstroms	10^6
Angstrom units	Micromicrons	10^2
Angstrom units	Microns	10^{-4}
Angstrom units	Milliangstroms	10^3
Angstrom units	Millimeters	10^{-7}
Angstrom units	Millimicrons	0.1
Ares	Acres	0.0247
Ares	Hectares	0.01
Ares	Sq decameters	1
Ares	Sq feet	1076.4
Ares	Sq meters	100
Ares	Sq miles	3.861×10^{-5}
Ares	Sq yards	119.6
Astronomical units	kilometers	1.495×10^8
Astronomical units	Light years	1.5804×10^{-5}
Astronomical units	Miles	9.29×10^8
Astronomical units	Parsecs	2.0879×10^5
Atmospheres	Bars	1.01325
Atmospheres	Dynes/sq cm	1.01325×10^6

To Convert From	To	Multiply By
Atmospheres	Grams/sq cm	1.03323×10^3
Atmospheres	Kilogram/sq meter	1.03323×10^4
Atmospheres	Pounds/sq ft	2116.2
Atmospheres	Pounds/sq in	14.696
Attoamperes	Amperes	10^{-18}
Attoamperes	Coulombs/sec	10^{-18}
Attoamperes	Femtoamperes	10^{-3}
Attoamperes	Kiloamperes	10^{-21}
Attoamperes	Megamperes	10^{-24}
Attoamperes	Microamperes	10^{-12}
Attoamperes	Milliamperes	10^{-15}
Attoamperes	Nanoamperes	10^{-9}
Attoamperes	Picoamperes	10^{-6}
Attofarads	Farads	10^{-18}
Attofarads	Femtofarads	10^{-3}
Attofarads	Microfarads	10^{-12}
Attofarads	Micromicrofarads	10^{-6}
Attofarads	Nanofarads	10^{-9}
Attofarads	Picofarads	10^{-6}
Attovolts	Femtovolts	10^{-3}
Attovolts	Gigavolts	10^{-27}
Attovolts	Kilovolts	10^{-21}
Attovolts	Megavolts	10^{-24}
Attovolts	Microvolts	10^{-12}
Attovolts	Millivolts	10^{-15}
Attovolts	Nanovolts	10^{-9}
Attovolts	Picovolts	10^{-6}
Attovolts	Teravolts	10^{-30}
Attovolts	Volts	10^{-18}
Attowatts	Femtowatts	10^{-3}
Attowatts	Gigawatts	10^{-27}
Attowatts	Horsepower	1.34×10^{-21}
Attowatts	Kilowatts	10^{-21}
Attowatts	Megawatts	10^{-24}
Attowatts	Microwatts	10^{-12}
Attowatts	Milliwatts	10^{-15}
Attowatts	Nanowatts	10^{-9}
Attowatts	Picowatts	10^{-6}
Attowatts	Terawatts	10^{-30}
Attowatts	Watts	10^{-18}

To Convert From	To	Multiply By
Average value (sine wave)	Effective value	1.112
Average value (sine wave)	Maximum value	1.57
Average value (sine wave)	RMS value	1.112
Average value (square wave)	Effective value	1
Average value (square wave)	Maximum value	1
Average value (square wave)	RMS value	1
Bars	Atmospheres	0.9869
Bars	Dynes/sq cm	10^6
Bars	Grams/sq cm	1019.72
Bars	Kilogram/sq cm	1.01972
Bars	Kilogram/sq meter	1.01972×10^4
Bars	Microbars	10^6
Bars	Millibars	10^3
Bars	Millitorrs	7.5×10^5
Bars	Pounds/sq ft	2088.55
Bars	Pounds/sq in.	14.504
Bars	Torrs	750
Bels	Decibels	10
Btu (mean)	Ergs	1.0549×10^{10}
Btu (mean)	Foot-poundals	2.50323×10^4
Btu (mean)	Foot-pounds	778.03
Btu (mean)	Gram-cm	1.0757×10^7
Btu (mean)	Hp-hr	3.93×10^{-4}
Btu (mean)	Joules (Int)	1054.7
Btu (mean)	Kilogram-meters	107.57
Btu (mean)	Kilowatt-hr (abs)	2.93×10^{-4}
Btu (mean)	Watt-hr (abs)	0.29302
Btu (mean)	Watt-sec (abs)	1054.9
Candles/sq cm	Foot lamberts	2918
Candles/sq cm	Lamberts	3.1416
Candles/sq cm	Millilamberts	3141.6
Candles/sq ft	Foot lamberts	3.1416
Candles/sq ft	Lamberts	3.38×10^{-3}
Candles/sq ft	Millilamberts	3.382

To Convert From	To	Multiply By
Candles/sq in.	Foot lamberts	452.4
Candles/sq in.	Lamberts	0.4869
Candles/sq in.	Millilamberts	486.9
Centares	Ares	0.01
Centares	Sq feet	10.764
Centares	Sq inches	1550
Centares	Sq meters	1
Centares	Sq yards	1.196
Centigrams	Grams	0.01
Centigrams	Kilograms	10^{-5}
Centigrams	Milligrams	10
Centiliters	Cubic cm	10
Centiliters	Cubic in.	0.61025
Centiliters	Fluid ounces	0.33815
Centiliters	Liters	0.01
Centimeters	Angstrom units	10^8
Centimeters	Feet	0.03281
Centimeters	Inches	0.3937
Centimeters	Kilometers	10^{-5}
Centimeters	Meters	0.01
Centimeters	Microns	10^4
Centimeters	Miles	6.214×10^{-6}
Centimeters	Millimeters	10
Centimeters	Mils	393.7
Centimeters	Yards	0.01094
Centimeters/sec	Feet/min	1.9685
Centimeters/sec	Feet/sec	0.032808
Centimeters/sec	Kilometer/hr	0.036
Centimeters/sec	Kilometer/min	6×10^{-4}
Centimeters/sec	Meters/min	0.6
Centimeters/sec	Meters/sec	0.01
Centimeters/sec	Miles/hr	0.0223693
Centimeters/sec	Miles/min	3.73×10^{-4}
Centimeters/sec	Miles/sec	6.2137×10^{-6}
Centipoises	Poises	0.01
Centistokes	Stokes	0.01
Circles	Circumferences	1
Circles	Degrees	360
Circles	Quadrants	4
Circles	Radians	6.283

To Convert From	To	Multiply By
Circular inches	Circ mils	1×10^6
Circular inches	Sq cm	5.0671
Circular inches	Sq in.	0.7854
Circular inches	Sq mils	7.854×10^5
Circular mils	Sq cm	5.0671×10^{-6}
Circular mils	Sq in.	7.854×10^{-7}
Circular mils	Sq mils	0.7854
Circular mils	Sq mm	5.0671×10^{-4}
Circular mm	Sq cm	7.854×10^{-3}
Circular mm	Sq mm	0.7854
Circumferences	Circles	1
Circumferences	Degrees	360
Circumferences	Grads	400
Circumferences	Minutes	21,600
Circumferences	Quadrants	4
Circumferences	Radians	6.283
Circumferences	Revolutions	1
Circumferences	Seconds	1296
Common logarithm	Natural log	2.303
Coulombs	Abcoulombs	0.1
Coulombs	Faradays	1.03×10^{-5}
Coulombs/sec	Amperes	1
Coulombs/sec	Attoamperes	10^{18}
Coulombs/sec	Femtoamperes	10^{15}
Coulombs/sec	Kiloamperes	10^{-3}
Coulombs/sec	Megamperes	10^{-6}
Coulombs/sec	Microamperes	10^6
Coulombs/sec	Milliamperes	10^3
Coulombs/sec	Nanoamperes	10^9
Coulombs/sec	Picoamperes	10^{12}
Cubic cm	Cu decameters	10^{-9}
Cubic cm	Cu decimeters	10^{-3}
Cubic cm	Cu feet	3.53144×10^{-5}
Cubic cm	Cu inches	0.0610234
Cubic cm	Cu meters	10^{-6}
Cubic cm	Cu mm	1000
Cubic cm	Fluid ounces	0.033814
Cubic cm	Gallons (liq)	2.642×10^{-4}
Cubic cm	Liters	10^{-3}
Cubic cm	Cu milliliters	1

To Convert From	To	Multiply By
Cubic cm	Pints (liq)	2.1134×10^{-3}
Cubic cm	Quarts (liq)	1.0567×10^{-3}
Cubic ft	Cu cm	2.8317×10^4
Cubic ft	Cu inches	1728
Cubic ft	Cu meters	0.028317
Cubic ft	Cu yards	0.037037
Cubic feet	Fluid ounces	957.51
Cubic feet	Gallons (liq)	7.48052
Cubic feet	Kiloliters	0.028316
Cubic feet	Liters	28.316
Cubic feet	Pints (liq)	59.844
Cubic feet	Quarts (liq)	29.922
Cubic feet/min	Cubic cm/sec	471.95
Cubic feet/min	Cubic ft/sec	0.016667
Cubic inches	Centiliters	1.63867
Cubic inches	Cu cm	16.3872
Cubic inches	Cu decimeters	0.0163872
Cubic inches	Cu feet	5.787×10^{-4}
Cubic inches	Cu meters	1.63872×10^{-5}
Cubic inches	Cu yards	2.14335×10^{-5}
Cubic inches	Decaliters	1.6386×10^{-3}
Cubic inches	Fluid ounces	0.55411
Cubic inches	Gallons (liq)	4.329×10^{-3}
Cubic inches	Liters	0.0163867
Cubic inches	Milliliters	16.3867
Cubic inches	Pints (liq)	0.03462
Cubic inches	Quarts (liq)	0.017316
Cubic meters	Cu cm	10^6
Cubic meters	Cu decimeters	10^3
Cubic meters	Cu feet	35.3144
Cubic meters	Cu hectometers	10^{-6}
Cubic meters	Cu inches	6.1023×10^4
Cubic meters	Cu kilometers	10^{-9}
Cubic meters	Cu mm	10^9
Cubic meters	Cu yards	1.30794
Cubic meters	Gallons (liq)	264.173
Cubic meters	Kiloliters	1
Cubic meters	Liters	10^3
Cubic meters	Pints (liq)	2113.4
Cubic meters	Quarts (liq)	1056.7

To Convert From	To	Multiply By
Cubic millimeters	Cu cm	10^{-3}
Cubic millimeters	Cu inches	6.1023×10^{-5}
Cubic millimeters	Cu meters	10^{-9}
Curies	Kilocuries	10^{-3}
Curies	Megacuries	10^{-6}
Curies	Microcuries	10^{6}
Curies	Millicuries	10^{3}
Curies	Rutherfords	3.71×10^{4}
Cycles	Gigacycles	10^{-9}
Cycles	Kilocycles	10^{-3}
Cycles	Megacycles	10^{-6}
Cycles	Teracycles	10^{-12}
Cycles/cm	Cycles/ft	30.48
Cycles/cm	Cycles/in.	2.54
Cycles/ft	Cycles/cm	0.0328
Cycles/ft	Cycles/in.	0.08333
Cycles/hr	Cycles/min	0.01667
Cycles/hr	Cycles/sec	2.777×10^{-4}
Cycles/in.	Cycles/cm	0.3937
Cycles/in.	Cycles/ft	12
Cycles/min	Cycles/hr	60
Cycles/min	Cycles/sec	0.01667
Cycles/sec	Cycles/hr	3600
Cycles/sec	Cycles/min	60
Cycles/sec	Gigacycles/sec	10^{-9}
Cycles/sec	Gigahertz	10^{-9}
Cycles/sec	Hertz	1
Cycles/sec	Kilocycles/sec	10^{-3}
Cycles/sec	Kilohertz	10^{-3}
Cycles/sec	Megacycles/sec	10^{-6}
Cycles/sec	Megahertz	10^{-6}
Cycles/sec	Teracycles/sec	10^{-12}
Cycles/sec	Terahertz	10^{-12}
Days	Hours	24
Days	Microseconds	8.64×10^{10}
Days	Milliseconds	8.64×10^{7}
Days	Minutes	1440
Days	Nanoseconds	8.64×10^{13}
Days	Picoseconds	8.64×10^{16}
Days	Seconds	8.640×10^{4}

To Convert From	To	Multiply By
Decagrams	Grams	10
Decaliters	Cubic in.	610.25
Decaliters	Liters	10
Decameters	Meters	10
Decibels	Bels	0.1
Decigrams	Grams	0.1
Decigrams	Kilograms	10^{-4}
Decigrams	Milligrams	100
Deciliters	Kiloliters	10^{-4}
Deciliters	Liters	0.1
Deciliters	Milliliters	100
Decimeters	Kilometers	10^{-4}
Decimeters	Meters	0.1
Decimeters	Millimeters	100
Degrees (angle)	Circumferences	2.78×10^{-3}
Degrees (angle)	Grads	1.111
Degrees (angle)	Minutes	60
Degrees (angle)	Quadrants	0.0111
Degrees (angle)	Radians	0.0174533
Degrees (angles)	Revolutions	2.778×10^{-3}
Degrees (angle)	Right angles	0.0111
Degrees (angle)	Seconds	3600
Degrees (angle)/sec	Radians/sec	0.0174533
Dynes	Grams	1.019×10^{-3}
Dynes	Kilograms	1.0197×10^{-6}
Dynes	Poundals	7.233×10^{-5}
Dynes	Pounds	2.248×10^{-6}
Dyne-cm	Ergs	1
Dyne-cm	Foot-poundals	2.373×10^{-6}
Dyne-cm	Foot-pounds	7.375×10^{-8}
Dyne-cm	Gram-cm	1.0197×10^{-3}
Dyne-cm	Inch-lb	8.851×10^{-7}
Dyne-cm	Kilogram-meters	1.01972×10^{-8}
Effective value (sine wave)	Average value	0.899
Effective value (sine wave)	Maximum value	1.414
Effective value (square wave)	Average value	1
Effective value (square wave)	Maximum value	1

To Convert From	To	Multiply By
Ergs	Btu (mean)	9.479×10^{-11}
Ergs	Dyne-cm	1
Ergs	Foot-poundals	2.373×10^{-6}
Ergs	Foot-pounds	7.375×10^{-8}
Ergs	Gram-cm	1.1019×10^{-3}
Ergs	Joules (abs)	10^{-7}
Ergs	Kilogram-meters	1.0197×10^{-8}
Erg-sec	Planck's constants	1.5097×10^{26}
Faradays	Abcoulombs	10^4
Faradays	Ampere-hours	26.8
Faradays	Coulombs	9.6494×10^4
Faradays	Milliampere-hours	0.0268
Farads	Attofarads	10^{18}
Farads	Femtofarads	10^{15}
Farads	Microfarads	10^6
Farads	Micromicrofarads	10^{12}
Farads	Nanofarads	10^9
Farads	Picofarads	10^{12}
Feet	Centimeters	30.48
Feet	Inches	12
Feet	Kilometers	3.048×10^{-4}
Feet	Meters	0.3048
Feet	Miles	1.894×10^{-4}
Feet	Millimeters	304.8
Feet	Mils	1.2000×10^4
Feet	Yards	0.3333
Feet/hr	Cm/hr	30.48
Feet/hr	Cm/min	0.5080
Feet/hr	Cm/sec	8.467×10^{-3}
Feet/hr	Feet/min	0.01667
Feet/hr	Feet/sec	2.778×10^{-4}
Feet/hr	Kilometers/hr	3.048×10^{-4}
Feet/hr	Kilometers/min	5.08×10^{-6}
Feet/hr	Kilometers/sec	8.467×10^{-8}
Feet/hr	Meters/hr	0.3048
Feet/hr	Meters/min	5.08×10^{-3}
Feet/hr	Meters/sec	9.467×10^{-5}
Feet/hr	Miles/hr	1.894×10^{-4}
Feet/hr	Miles/min	3.1566×10^{-6}
Feet/hr	Miles/sec	5.261×10^{-8}

To Convert From	To	Multiply By
Feet/min	Cm/sec	0.508
Feet/min	Feet/hr	60
Feet/min	Feet/sec	0.01667
Feet/min	Kilometers/hr	0.01829
Feet/min	Kilometers/min	3.048×10^{-4}
Feet/min	Meters/min	0.3048
Feet/min	Meters/sec	5.08×10^{-3}
Feet/min	Miles/hr	0.011364
Feet/min	Miles/min	1.894×10^{-4}
Feet/min	Miles/sec	1.1364×10^{-3}
Feet/sec	Feet/hr	3600
Feet/sec	Feet/min	60
Feet/sec	Kilometer/hr	1.0973
Feet/sec	Kilometer/min	0.01829
Feet/sec	Mach number	8.857×10^{-4}
Feet/sec	Meters/min	18.288
Feet/sec	Meters/sec	0.3048
Feet/sec	Miles/hr	0.6818
Feet/sec	Miles/min	0.011364
Feet/sec	Miles/sec	1.894×10^{-3}
Femtoamperes	Amperes	10^{-15}
Femtoamperes	Attoamperes	10^{3}
Femtoamperes	Coulombs/sec	10^{-15}
Femtoamperes	Kiloamperes	10^{-18}
Femtoamperes	Megamperes	10^{-21}
Femtoamperes	Microamperes	10^{-9}
Femtoamperes	Milliamperes	10^{-12}
Femtoamperes	Nanoamperes	10^{-6}
Femtoamperes	Picoamperes	10^{-3}
Femtofarads	Attofarads	10^{3}
Femtofarads	Farads	10^{-15}
Femtofarads	Microfarads	10^{-9}
Femtofarads	Micromicrofarads	10^{-3}
Femtofarads	Nanofarads	10^{-6}
Femtofarads	Picofarads	10^{-3}
Femtovolts	Attovolts	10^{3}
Femtovolts	Gigavolts	10^{-24}
Femtovolts	Kilovolts	10^{-18}
Femtovolts	Megavolts	10^{-21}
Femtovolts	Microvolts	10^{-9}

To Convert From	To	Multiply By
Femtovolts	Millivolts	10^{-12}
Femtovolts	Nanovolts	10^{-6}
Femtovolts	Picovolts	10^{-3}
Femtovolts	Teravolts	10^{-27}
Femtovolts	Volts	10^{-15}
Femtowatts	Attowatts	10^3
Femtowatts	Gigawatts	10^{-24}
Femtowatts	Horsepower	1.34×10^{-18}
Femtowatts	Kilowatts	10^{-18}
Femtowatts	Megawatts	10^{-21}
Femtowatts	Microwatts	10^{-9}
Femtowatts	Milliwatts	10^{-12}
Femtowatts	Nanowatts	10^{-6}
Femtowatts	Picowatts	10^{-3}
Femtowatts	Terawatts	10^{-27}
Femtowatts	Watts	10^{-15}
Fluid ounces	Cubic cm	29.5737
Fluid ounces	Cubic ft	1.045×10^{-3}
Fluid ounces	Cubic in.	1.8047
Fluid ounces	Cubic meters	2.957×10^{-5}
Fluid ounces	Cubic yards	3.868×10^{-5}
Fluid ounces	Gallons (liq)	7.813×10^{-3}
Fluid ounces	Gills	0.25
Fluid ounces	Liters	0.029573
Fluid ounces	Milliliters	29.573
Fluid ounces	Pints (liq)	0.0625
Fluid ounces	Quarts (liq)	0.03125
Foot candles	Lumens/sq meter	10.7639
Foot candles	Lumens/sq ft	1
Foot candles	Lux	10.764
Foot candles	Milliphots	1.0764
Foot candles	Phots	1.076×10^{-3}
Foot-lamberts	Candles/sq cm	3.426×10^{-4}
Foot-lamberts	Candles/sq in.	2.2×10^{-3}
Foot-lamberts	Candles/sq ft	0.3183
Foot-lamberts	Lamberts	1.1×10^{-3}
Foot-poundals	Dyne-cm	4.21402×10^5
Foot-poundals	Gram-cm	429.711
Foot-poundals	Kilogram-meters	4.2971×10^{-3}
Foot-pounds	Dyne-cm	1.35582×10^7

To Convert From	To	Multiply By
Foot-pounds	Gram-cm	1.38255×10^4
Foot-pounds	Kilogram-meters	0.138255
Gallons (liq)	Cubic cm	3785
Gallons (liq)	Cu meters	3.785×10^{-3}
Gallons (liq)	Kiloliters	3.785×10^{-3}
Gallons (liq)	Liters	3.785
Gallons (liq)	Milliliters	3785
Gausses (abs)	Gausses (int)	0.99966
Gausses (abs)	Lines/sq cm	1
Gausses (abs)	Lines/sq in.	6.4516
Gausses (abs)	Maxwells/sq cm	1
Gausses (abs)	Maxwells/sq in.	0.4516
Gausses (abs)	Webers/sq cm	10^{-8}
Gausses (abs)	Webers/sq in.	6.4516×10^{-8}
Gausses (int)	Gausses (abs)	1.00034
Gigacycles	Cycles	10^9
Gigacycles	Kilocycles	10^6
Gigacycles	Megacycles	10^3
Gigacycles	Teracycles	10^{-3}
Gigacycles/sec	Cycles/sec	10^9
Gigacycles/sec	Gigahertz	1
Gigacycles/sec	Hertz	10^9
Gigacycles/sec	Kilocycles/sec	10^6
Gigacycles/sec	Kilohertz	10^6
Gigacycles/sec	Megacycles/sec	10^3
Gigacycles/sec	Megahertz	10^3
Gigacycles/sec	Teracycles/sec	10^{-3}
Gigacycles/sec	Terahertz	10^{-3}
Gigahertz	Cycles/sec	10^9
Gigahertz	Gigacycles/sec	1
Gigahertz	Hertz	10^9
Gigahertz	Kilocycles/sec	10^6
Gigahertz	Kilohertz	10^6
Gigahertz	Megacycles/sec	10^3
Gigahertz	Megahertz	10^3
Gigahertz	Teracycles/sec	10^{-3}
Gigahertz	Terahertz	10^{-3}
Gigavolts	Attovolts	10^{27}
Gigavolts	Femtovolts	10^{24}
Gigavolts	Kilovolts	10^6

To Convert From	To	Multiply By
Gigavolts	Megavolts	10^3
Gigavolts	Microvolts	10^{15}
Gigavolts	Millivolts	10^{12}
Gigavolts	Nanovolts	10^{18}
Gigavolts	Picovolts	10^{21}
Gigavolts	Teravolts	10^{-3}
Gigavolts	Volts	10^9
Gigawatts	Attowatts	10^{27}
Gigawatts	Femtowatts	10^{24}
Gigawatts	Horsepower	1.34×10^6
Gigawatts	Kilowatts	10^6
Gigawatts	Megawatts	10^3
Gigawatts	Microwatts	10^{15}
Gigawatts	Milliwatts	10^{12}
Gigawatts	Nanowatts	10^{18}
Gigawatts	Picowatts	10^{21}
Gigawatts	Terawatts	10^{-3}
Gigawatts	Watts	10^9
Gigohms	Kilohms	10^6
Gigohms	Megohms	10^5
Gigohms	Microhms	10^{15}
Gigohms	Milliohms	10^{12}
Gigohms	Ohms	10^9
Gigohms	Teraohms	10^{-3}
Gigohms/cm	Gigohms/in.	2.54
Gigohms/cm	Gigohms/mil	2.54×10^{-3}
Gigohms/in.	Gigohms/cm	0.3937
Gigohms/in.	Gigohms/mil	10^{-3}
Gigohms/mil	Gigohms/cm	393.7
Gigohms/mil	Gigohms/in.	10^3
Gilberts (abs)	Abampere-turns	0.0795
Gilberts (abs)	Ampere-turns	0.7958
Gilberts (abs)	Gilberts (int)	1.00014
Gilberts (int)	Gilberts (abs)	0.99986
Gills	Cubic cm	118.29
Gills	Gallons (liq)	0.03125
Gills	Liters	0.11829
Gills	Milliliters	118.29
Gills	Pints (liq)	0.25
Gills	Quarts (liq)	0.125

To Convert From	To	Multiply By
Grads	Circumferences	2.5×10^{-3}
Grads	Degrees	0.9
Grads	Minutes	54
Grads	Quadrants	0.01
Grads	Radians	0.01571
Grads	Right angles	0.01
Grads	Seconds	3240
Grads/sec	Radians/sec	0.157
Grams	Decagrams	0.1
Grams	Decigrams	10
Grams	Dynes	980.665
Grams	Hectograms	10^{-2}
Grams	Kilograms	10^{-3}
Grams	Micrograms	10^6
Grams	Milligrams	10^3
Grams	Myriagrams	10^{-4}
Grams	Ounces (avdp)	0.035274
Grams	Poundals	0.0709315
Grams	Pounds (avdp)	2.204×10^{-3}
Grams	Tons (long)	9.842×10^{-7}
Grams	Tons (metric)	10^{-6}
Grams	Tons (short)	1.1023×10^{-6}
Grams/cm	Dynes/cm	980.665
Grams/cm	Poundals/in.	0.180166
Grams/cm	Pounds/ft	0.067197
Grams/cm	Pounds/in.	5.5997×10^{-3}
Grams/cm	Pounds/mi.	354.8
Grams/cm	Pounds/yard	0.201591
Grams/cu cm	Dynes/cu cm	980.665
Grams/cu cm	Grams/milliliter	1
Grams/cu cm	Kilogram/cu meter	10^3
Grams/cu cm	Kilogram/hectoliter	100
Grams/cu cm	Kilogram/liter	1
Grams/cu cm	Pounds/cu ft	62.4282
Grams/cu cm	Pounds/cu in.	0.03613
Grams/cu cm	Pounds/cu yard	1685.56
Grams/liter	Grams/cu cm	10^{-3}
Grams/liter	Pounds/cu ft	0.062427
Grams/milliliter	Grams/cu cm	1
Grams/sq cm	Atmospheres	9.678×10^{-4}

To Convert From	To	Multiply By
Grams/sq cm	Bars	9.08×10^{-4}
Grams/sq cm	Dynes/sq cm	980.665
Grams/sq cm	Kilograms/sq meter	10
Grams/sq cm	Poundals/sq in.	0.45762
Grams/sq cm	Pounds/sq ft	2.04816
Grams/sq cm	Pounds/sq in.	0.014223
Gram-cm	Ergs	980.665
Gram-cm	Foot-poundals	2.327×10^{-3}
Gram-cm	Foot-pounds	7.233×10^{-5}
Gram-cm	Kilogram-meters	10^{-5}
Hectares	Acres	2.471
Hectares	Ares	10^2
Hectares	Centares	10^4
Hectares	Sq cm	10^8
Hectares	Sq feet	1.07139×10^5
Hectares	Sq inches	1.55×10^7
Hectares	Sq km	10^{-2}
Hectares	Sq meters	10^4
Hectares	Sq miles	3.861×10^{-3}
Hectares	Sq yards	1.19598×10^4
Hectograms	Dynes	9.80665×10^4
Hectograms	Grams	100
Hectograms	Kilograms	0.1
Hectograms	Ounces (avdp)	3.5274
Hectograms	Poundals	7.09315
Hectograms	Pounds (avdp)	0.22046
Hectoliters	Bushels	2.83782
Hectoliters	Cu cm	10^5
Hectoliters	Cu decimeters	100
Hectoliters	Cu ft	3.53154
Hectoliters	Cu in.	6102.5
Hectoliters	Cu meters	0.1
Hectoliters	Cu yards	0.130798
Hectoliters	Decaliters	10
Hectoliters	Gallons (liq)	26.418
Hectoliters	Gills	845.37
Hectoliters	Kiloliters	0.1
Hectoliters	Liters	100
Hectoliters	Microliters	10^8
Hectoliters	Milliliters	10^5

To Convert From	To	Multiply By
Hectoliters	Pints (liq)	311.342
Hectoliters	Quarts (liq)	105.671
Hectometers	Centimeters	10^4
Hectometers	Feet.	328.08
Hectometers	Inches	3937
Hectometers	Kilometers	0.1
Hectometers	Megameters	10^{-4}
Hectometers	Meters	100
Hectometers	Microns	10^8
Hectometers	Miles	0.062137
Hectometers	Millimeters	10^5
Hectometers	Millimicrons	10^{11}
Hectometers	Myriameters	0.01
Hectometers	Yards	109.361
Henrys	Abhenrys	10^9
Henrys	Microhenrys	10^6
Henrys	Millihenrys	10^3
Henrys	Nanohenrys	10^9
Henrys	Picohenrys	10^{12}
Hertz	Cycles/sec	1
Hertz	Gigacycles/sec	10^{-9}
Hertz	Gigahertz	10^{-9}
Hertz	Kilocycles/sec	10^{-3}
Hertz	Kilohertz	10^{-3}
Hertz	Megacycles/sec	10^{-6}
Hertz	Megahertz	10^{-6}
Hertz	Teracycles/sec	10^{-12}
Hertz	Terahertz	10^{-12}
Horsepower	Attowatts	7.46×10^{20}
Horsepower	Femtowatts	7.46×10^{17}
Horsepower	Gigawatts	7.46×10^{-7}
Horsepower	Kilowatts	0.746
Horsepower	Megawatts	7.46×10^{-4}
Horsepower	Microwatts	7.46×10^8
Horsepower	Milliwatts	7.46×10^5
Horsepower	Nanowatts	7.46×10^{11}
Horsepower	Picowatts	7.46×10^{14}
Horsepower	Terawatts	7.46×10^{-10}
Horsepower	Watts	746
Horsepower-hours	Kilowatt-hours	0.7455

215

To Convert From	To	Multiply By
Horsepower-hours	Watt-seconds	2.684×10^6
Hours	Days	0.04167
Hours	Microseconds	3.6×10^9
Hours	Milliseconds	3.6×10^6
Hours	Minutes	60
Hours	Nanoseconds	3.6×10^{12}
Hours	Picoseconds	3.6×10^{15}
Hours	Seconds	3600
Inches	Centimeters	2.54
Inches	Feet	0.0833333
Inches	Kilometers	2.54×10^{-5}
Inches	Meters	0.0254
Inches	Microns	2.5400×10^4
Inches	Miles	6.3360×10^4
Inches	Millimeters	25.4
Inches	Millimicrons	2.54×10^7
Inches	Mils	10^3
Inches	Yards	0.27777
Joules	Btu	9.48×10^{-4}
Joules	Dyne-cm	10^7
Joules	Ergs	10^7
Joules	Foot-poundals	23.735
Joules	Foot-pounds	0.737
Joules	Gram-cm	1.0199×10^4
Joules	Hp-hr	3.726×10^{-7}
Joules	Kilogram-meters	0.1019
Joules	Kilowatt-hr	2.778×10^{-7}
Joules	Watt-hr	2.778×10^{-4}
Joules	Watt-sec	1
Kiloamperes	Amperes	10^3
Kiloamperes	Attoamperes	10^{21}
Kiloamperes	Coulombs/sec	10^3
Kiloamperes	Femtoamperes	10^{18}
Kiloamperes	Megamperes	10^{-3}
Kiloamperes	Microamperes	10^9
Kiloamperes	Milliamperes	10^6
Kiloamperes	Nanoamperes	10^{12}
Kiloamperes	Picoamperes	10^{15}
Kilocuries	Curies	10^3

To Convert From	To	Multiply By
Kilocuries	Megacuries	10^{-3}
Kilocuries	Microcuries	10^9
Kilocuries	Millicuries	10^6
Kilocycles	Cycles	10^3
Kilocycles	Gigacycles	10^{-6}
Kilocycles	Megacycles	10^{-3}
Kilocycles	Teracycles	10^{-9}
Kilocylces/sec	Cycles/sec	10^3
Kilocycles/sec	Gigacycles/sec	10^{-6}
Kilocycles/sec	Gigahertz	10^{-6}
Kilocycles/sec	Hertz	10^3
Kilocycles/sec	Kilohertz	1
Kilocycles/sec	Megacycles/sec	10^{-3}
Kilocycles/sec	Megahertz	10^{-3}
Kilocycles/sec	Teracycles/sec	10^{-9}
Kilocycles/sec	Terahertz	10^{-9}
Kilograms	Decagrams	100
Kilograms	Dynes	9.80665×10^5
Kilograms	Grams	10^3
Kilograms	Milligrams	10^6
Kilograms	Myriagrams	0.1
Kilograms	Ounces (avdp)	35.274
Kilograms	Poundals	70.931
Kilograms	Pounds (advp)	2.2046
Kilograms	Quintals	0.01
Kilograms	Tons (long)	9.84×10^{-4}
Kilograms	Tons (metric)	10^{-3}
Kilograms	Tons (short)	1.102×10^{-3}
Kilograms/cu meter	Grams/cu cm	10^{-3}
Kilograms/cu meter	Kilogram/liter	10^{-3}
Kilograms/cu meter	Lb/cu ft	0.06243
Kilograms/cu meter	Lb/cu in.	3.61275×10^{-5}
Kilograms/cu meter	Lb/cu yard	1.6856
Kilograms/sq cm	Atmospheres	0.96784
Kilograms/sq cm	Bars	0.980665
Kilograms/sq cm	Dynes/sq cm	9.80665×10^5
Kilograms/sq cm	Grams/sq cm	1000
Kilograms/sq cm	Pounds/sq ft	2048.15
Kilograms/sq cm	Pound/sq in.	14.223

To Convert From	To	Multiply By
Kilograms/sq cm	Tons (short)/sq ft	1.0241
Kilograms/sq cm	Tons (short)/sq in.	7.11×10^{-3}
Kilograms/sq meter	Atmospheres	9.678×10^{-5}
Kilograms/sq meter	Bars	9.80665×10^{-5}
Kilograms/sq meter	Dynes/sq cm	98.0665
Kilograms/sq meter	Grams/sq cm	0.1
Kilograms/sq meter	Pounds/sq ft	0.204815
Kilograms/sq meter	Pounds/sq in.	1.42×10^{-3}
Kilograms/sq meter	Tons (short)/sq ft	1.024×10^{-4}
Kilograms/sq meter	Tons (short)/sq in.	7.11165×10^{-7}
Kilograms/sq mm	Atmospheres	96.7841
Kilograms/sq mm	Bars	98.0665
Kilograms/sq mm	Dynes/sq cm	9.80665×10^7
Kilograms/sq mm	Grams/sq cm	10^5
Kilograms/sq mm	Pounds/sq ft	2.04815×10^5
Kilograms/sq mm	Pounds/sq in.	1422.33
Kilogram-meters	Dyne-cm	9.80665×10^7
Kilogram-meters	Ergs	9.80665×10^7
Kilogram-meters	Foot-poundals	232.715
Kilogram-meters	Foot-pounds	7.233
Kilogram-meters	Gram-cm	10^5
Kilogram-meters	Hp-hr (metric)	3.7037×10^{-6}
Kilogram-meters	Hp-hr (U.S.)	3.65303×10^{-6}
Kilohertz	Cycles/sec	10^3
Kilohertz	Gigacycles/sec	10^{-6}
Kilohertz	Gigahertz	10^{-6}
Kilohertz	Hertz	10^3
Kilohertz	Kilocycles/sec	1
Kilohertz	Megacycles/sec	10^{-3}
Kilohertz	Megahertz	10^{-3}
Kilohertz	Teracycles/sec	10^{-9}
Kilohertz	Terahertz	10^{-9}
Kilohms	Gigohms	10^{-6}
Kilohms	Megohms	10^{-3}
Kilohms	Microhms	10^9
Kilohms	Milliohms	10^6
Kilohms	Ohms	10^3
Kilohms	Teraohms	10^{-9}
Kilohms/cm	Kilohms/in.	2.54
Kilohms/cm	Kilohms/mil	2.54×10^{-3}

To Convert From	To	Multiply By
Kilohms/in.	Kilohms/cm	0.3937
Kilohms/in.	Kilohms/mil	10^{-3}
Kilohms/mil	Kilohms/cm	393.7
Kilohms/mil	Kilohms/in.	10^3
Kilohms/volt	Megohms/volt	10^{-3}
Kilohms/volt	Ohms/volt	10^3
Kilolines	Lines	10^3
Kilolines	Maxwells	10^3
Kilolines	Webers	10^{-5}
Kiloliters	Cubic cm	10^6
Kiloliters	Cubic ft	35.3154
Kiloliters	Cubic in.	6.1025×10^4
Kiloliters	Cubic meters	1
Kiloliters	Cubic yards	1.30798
Kiloliters	Decaliters	100
Kiloliters	Gallons (liq)	264.178
Kiloliters	Hectoliters	10
Kiloliters	Liters	10^3
Kiloliters	Milliliters	10^6
Kiloliters	Pints (liq)	2113.4
Kiloliters	Quarts (liq)	1056.7
Kilometers	Astronomical units	6.689×10^{-9}
Kilometers	Centimeters	10^5
Kilometers	Feet	3280.8
Kilometers	Hectometers	10
Kilometers	Inches	3.9370×10^4
Kilometers	Light years	1.057×10^{-13}
Kilometers	Megameters	10^{-3}
Kilometers	Meters	10^3
Kilometers	Miles	0.62137
Kilometers	Millimeters	10^6
Kilometers	Mils	3.937×10^7
Kilometers	Myriameters	0.1
Kilometers	Yards	1093.61
Kilometers/hr	Cm/sec	27.7778
Kilometers/hr	Feet/hr	3280.83
Kilometers/hr	Feet/min	54.681
Kilometers/hr	Feet/sec	0.91134
Kilometers/hr	Kilometer/min	0.0166667
Kilometers/hr	Kilometer/sec	2.778×10^{-4}

To Convert From	To	Multiply By
Kilometers/hr	Meters/hr	10^3
Kilometers/hr	Meters/min	16.6667
Kilometers/hr	Meters/sec	0.277778
Kilometers/hr	Miles/hr	0.62137
Kilometers/hr	Miles/min	0.010356
Kilometers/hr	Miles/sec	1.726×10^{-4}
Kilometers/min	Cm/sec	1666.67
Kilometers/min	Feet/min	3280.84
Kilometers/min	Feet/sec	54.681
Kilometers/min	Km/hr	60
Kilometers/min	Km/sec	0.016667
Kilometers/min	Meters/min	1000
Kilometers/min	Meters/sec	16.6667
Kilometers/min	Miles/hr	37.2822
Kilometers/min	Miles/min	0.62137
Kiloroentgens	Megaroentgens	10^{-3}
Kiloroentgens	Microroentgens	10^9
Kiloroentgens	Milliroentgens	10^6
Kiloroentgens	Roentgens	10^3
Kilorutherfords	Megarutherfords	10^{-3}
Kilorutherfords	Microrutherfords	10^9
Kilorutherfords	Millirutherfords	10^6
Kilorutherfords	Rutherfords	10^3
Kilovolt-amperes	Megavolt-amperes	10^{-3}
Kilovolt-amperes	Volt-amperes	10^3
Kilovolts	Attovolts	10^{21}
Kilovolts	Femtovolts	10^{18}
Kilovolts	Gigavolts	10^{-6}
Kilovolts	Megavolts	10^{-3}
Kilovolts	Microvolts	10^9
Kilovolts	Millivolts	10^6
Kilovolts	Nanovolts	10^{12}
Kilovolts	Picovolts	10^{15}
Kilovolts	Teravolts	10^{-9}
Kilovolts	Volts	10^3
Kilowatt-hours	Foot-pounds	2.6557×10^6
Kilowatt-hours	Horsepower-hours	1.3413
Kilowatt-hours	Joules	3.6×10^6
Kilowatt-hours	Kilogram-meters	3.67171×10^6
Kilowatt-hours	Watt-hours	1000

To Convert From	To	Multiply By
Kilowatt-hours	Watt-seconds	3.599×10^6
Kilowatts	Attowatts	10^{21}
Kilowatts	Femtowatts	10^{18}
Kilowatts	Gigawatts	10^{-6}
Kilowatts	Horsepower	1.34
Kilowatts	Megawatts	10^{-3}
Kilowatts	Microwatts	10^9
Kilowatts	Milliwatts	10^6
Kilowatts	Nanowatts	10^{12}
Kilowatts	Picowatts	10^{15}
Kilowatts	Terawatts	10^{-9}
Kilowatts	Watts	10^3
Lamberts	Candles/sq cm	0.31831
Lamberts	Candles/sq ft	295.72
Lamberts	Candles/sq in.	2.0536
Lamberts	Foot-lamberts	929
Lamberts	Lumens/sq cm	1
Lamberts	Lumens/sq ft	929
Lamberts	Millilamberts	10^3
Light years	Astronomical units	6.327×10^4
Light years	Kilometers	9.461×10^{12}
Light years	Miles	5.87812×10^{12}
Light years	Parsecs	0.3067
Lines	Kilolines	10^{-3}
Lines	Maxwells	1
Lines/sq cm	Gausses (abs)	1
Lines/sq in.	Gausses (abs)	0.155
Lines/sq in.	Webers/sq cm	1.55×10^{-9}
Lines/sq in.	Webers/sq in.	10^{-8}
Liters	Cubic cm	10^3
Liters	Cu decimeters	1
Liters	Cu feet	0.03531
Liters	Cu inches	61.025
Liters	Cu meters	10^{-3}
Liters	Cu yards	1.308×10^{-3}
Liters	Decaliters	0.1
Liters	Gallons (liq)	0.26418
Liters	Gills	8.4537
Liters	Hectoliters	10^{-2}
Liters	Kiloliters	10^{-3}

To Convert From	To	Multiply By
Liters	Microliters	10^6
Liters	Milliliters	10^3
Liters	Pints (liq)	2.1134
Liters	Quarts (liq)	1.0567
Log_e	Log_{10}	0.4343
Log_{10}	Log_e	2.303
Lumens	Watts	1.47×10^{-3}
Lumens/sq cm	Lamberts	1
Lumens/sq ft	Lamberts	1.1×10^{-3}
Lumens/sq meter	Foot candles	0.092903
Lumens/sq meter	Lumens/sq ft	0.092903
Lumens/sq meter	Lux	1
Lumens/sq meter	Meter candles	1
Lumens/sq meter	Phots	10^{-4}
Lux	Lumens/sq meter	1
Lux	Meter candles	1
Lux	Microlux	10^6
Lux	Millilux	10^3
Mach number	Feet/sec	1129
Maximum value (sine wave)	Average value	0.637
Maximum value (sine wave)	Effective value	0.707
Maximum value (sine wave)	RMS value	0.707
Maximum value (square wave)	Average value	1
Maximum value (square wave)	Effective value	1
Maximum value (square wave)	RMS value	1
Maxwells (abs)	Kilolines	10^{-3}
Maxwells (abs)	Lines	1
Maxwells (abs)	Maxwells (int)	0.99966
Maxwells (int)	Maxwells (abs)	1.00034
Maxwells (int)	Millimaxwells	10^3
Maxwells (int)	Webers	10^{-8}
Megacuries	Curies	10^6
Megacuries	Kilocuries	10^3
Megacuries	Microcuries	10^{12}
Megacuries	Millicuries	10^9

To Convert From	To	Multiply By
Megacycles	Cycles	10^6
Megacycles	Gigacycles	10^{-3}
Megacycles	Kilocycles	10^3
Megacycles	Teracycles	10^{-6}
Megacycles/sec	Cycles/sec	10^6
Megacycles/sec	Gigacycles/sec	10^{-3}
Megacycles/sec	Gigahertz	10^{-3}
Megacycles/sec	Hertz	10^6
Megacycles/sec	Kilohertz	10^3
Megacycles/sec	Megahertz	1
Megacycles/sec	Teracycles/sec	10^{-6}
Megacycles/sec	Terahertz	10^{-6}
Megahertz	Cycles/sec	10^6
Megahertz	Gigacycles/sec	10^{-3}
Megahertz	Gigahertz	10^{-3}
Megahertz	Hertz	10^6
Megahertz	Kilocycles/sec	10^3
Megahertz	Kilohertz	10^3
Megahertz	Megacycles/sec	1
Megahertz	Teracycles/sec	10^{-6}
Megahertz	Terahertz	10^{-6}
Megameters	Kilometers	10^3
Megameters	Meters	10^6
Megameters	Miles	621.37
Megamperes	Amperes	10^6
Megamperes	Attoamperes	10^{24}
Megamperes	Coulombs/sec	10^6
Megamperes	Femtoamperes	10^{21}
Megamperes	Kiloamperes	10^3
Megamperes	Microamperes	10^{12}
Megamperes	Milliamperes	10^9
Megamperes	Nanoamperes	10^{15}
Megamperes	Picoamperes	10^{18}
Megaroentgens	Kiloroentgens	10^3
Megaroentgens	Microroentgens	10^{12}
Megaroentgens	Milliroentgens	10^9
Megaroentgens	Roentgens	10^6
Magarutherfords	Kilorutherfords	10^3
Megarutherfords	Microrutherfords	10^{12}
Megarutherfords	Millirutherfords	10^9

To Convert From	To	Multiply By
Megarutherfords	Rutherfords	10^6
Megavolt-amperes	Kilovolt-amperes	10^3
Megavolt-amperes	Volt-amperes	10^6
Megavolts	Attovolts	10^{24}
Megavolts	Femtovolts	10^{21}
Megavolts	Gigavolts	10^{-3}
Megavolts	Kilovolts	10^3
Megavolts	Microvolts	10^{12}
Megavolts	Millivolts	10^9
Megavolts	Nanovolts	10^{15}
Megavolts	Picovolts	10^{18}
Megavolts	Teravolts	10^{-6}
Megavolts	Volts	10^6
Megawatts	Attowatts	10^{24}
Megawatts	Femtowatts	10^{21}
Megawatts	Gigawatts	10^{-3}
Megawatts	Horsepower	1340
Megawatts	Kilowatts	10^3
Megawatts	Microwatts	10^{12}
Megawatts	Milliwatts	10^9
Megawatts	Nanowatts	10^{15}
Megawatts	Picowatts	10^{18}
Megawatts	Terawatts	10^{-6}
Megawatts	Watts	10^6
Megmhos	Mhos	10^6
Megmhos	Micromhos	10^{12}
Megohms	Gigohms	10^{-3}
Megohms	Kilohms	10^3
Megohms	Microhms	10^{12}
Megohms	Milliohms	10^9
Megohms	Ohms	10^6
Megohms	Teraohms	10^{-6}
Megohms/cm	Megohms/in.	2.54
Megohms/cm	Megohms/mil	2.54×10^{-3}
Megohms/in.	Megohms/cm	0.3937
Megohms/in.	Megohms/mil	10^{-3}
Megohms/mil	Megohms/cm	393.7
Megohms/mil	Megohms/in.	10^3
Megohms/volt	Kilohms/volt	10^3
Megohms/volt	Ohms/volt	10^6

To Convert From	To	Multiply By
Meters	Centimeters	10^2
Meters	Feet	0.3281
Meters	Hectometers	10^{-2}
Meters	Inches	39.37
Meters	Kilometers	10^{-3}
Meters	Megameters	10^{-6}
Meters	Micromicrons	10^{12}
Meters	Microns	10^6
Meters	Miles	6.214×10^{-4}
Meters	Millimeters	10^3
Meters	Millimicrons	10^9
Meters	Mils	3.9370×10^4
Meters	Myriameters	10^{-4}
Meters	Yards	1.0936
Meters/hr	Centimeters/hr	100
Meters/hr	Centimeters/min	1.6667
Meters/hr	Centimeters/sec	0.027778
Meters/hr	Feet/hr	3.28083
Meters/hr	Feet/min	0.05468
Meters/hr	Feet/sec	9.11×10^{-4}
Meters/hr	Kilometers/hr	10^{-3}
Meters/hr	Kilometers/min	1.6667×10^{-5}
Meters/hr	Kilometers/sec	2.7778×10^{-7}
Meters/hr	Meter/min	0.01667
Meters/hr	Meters/sec	2.778×10^{-4}
Meters/hr	Miles/hr	6.214×10^{-4}
Meters/hr	Miles/min	1.03562×10^{-5}
Meters/hr	Miles/sec	1.726×10^{-7}
Meters/min	Cm/sec	1.6667
Meters/min	Feet/hr	196.85
Meters/min	Feet/min	3.2808
Meters/min	Feet/sec	0.05468
Meters/min	Kilometer/hr	0.06
Meters/min	Kilometer/min	10^{-3}
Meters/min	Meters/sec	0.01667
Meters/min	Miles/hr	0.03728
Meters/min	Miles/min	6.214×10^{-4}
Meters/sec	Cm/sec	100
Meters/sec	Feet/sec	3.281
Meters/sec	Kilometer/hr	3.6

To Convert From	To	Multiply By
Meters/sec	Kilometer/min	0.06
Meters/sec	Meters/min	60
Meters/sec	Miles/hr	2.2369
Meters/sec	Miles/min	0.03728
Mhos	Abmhos	10^{-9}
Mhos	Megmhos	10^{-6}
Mhos	Micromhos	10^{-6}
Microamperes	Amperes	10^{-6}
Microamperes	Attoamperes	10^{12}
Microamperes	Coulombs/sec	10^{-6}
Microamperes	Femtoamperes	10^{9}
Microamperes	Kiloamperes	10^{-9}
Microamperes	Megamperes	10^{-12}
Microamperes	Milliamperes	10^{-3}
Microamperes	Nanoamperes	10^{3}
Microamperes	Picoamperes	10^{6}
Microamperes/volt	Amperes/volt	10^{-6}
Microamperes/volt	Milliamperes/volt	10^{-3}
Microangstroms	Angstroms	10^{-6}
Microangstroms	Milliangstroms	10^{-3}
Microbars	Bars	10^{-6}
Microbars	Millibars	10^{-3}
Microbars	Millitorrs	0.075
Microbars	Torrs	7.5×10^{-4}
Microcuries	Curies	10^{-6}
Microcuries	Kilocuries	10^{-9}
Microcuries	Megacuries	10^{-12}
Microcuries	Millicuries	10^{-3}
Microfarads	Attofarads	10^{12}
Microfarads	Farads	10^{-6}
Microfarads	Femtofarads	10^{9}
Microfarads	Micromicrofarads	10^{6}
Microfarads	Nanofarads	10^{3}
Microfarads	Picofarads	10^{6}
Micrograms	Grams	10^{-6}
Micrograms	Milligrams	10^{-3}
Microhenrys	Henrys	10^{-6}
Microhenrys	Millihenrys	10^{-3}
Microhenrys	Nanohenrys	10^{3}
Microhenrys	Picohenrys	10^{6}

To Convert From	To	Multiply By
Microhm-cm	Circ mil-ohms/ft	6.0153
Microhm-cm	Microhm-in.	0.3937
Microhm-cm	Ohm-cm	10^{-6}
Microhm-in.	Circ mil-ohms/ft	15.279
Microhm-in.	Microhm-cm	2.54
Microhms	Gigohms	10^{-15}
Microhms	Kilohms	10^{-9}
Microhms	Megohms	10^{-12}
Microhms	Milliohms	10^{-3}
Microhms	Ohms	10^{-6}
Microhms	Teraohms	10^{-18}
Microhms/cm	Microhms/in.	2.54
Microhms/cm	Microhms/mil	2.54×10^{-3}
Microhms/in.	Microhms/cm	0.3937
Microhms/in.	Microhms/mil	10^{-3}
Microhms/mil	Microhms/cm	393.7
Microhoms/mil	Microhms/in.	10^3
Microliters	Cubic cm	10^3
Microliters	Liters	10^{-6}
Microliters	Milliliters	10^{-3}
Microlux	Lux	10^{-6}
Microlux	Millilux	10^{-3}
Micromhos	Abmhos	10^{-15}
Micromhos	Megmhos	10^{-12}
Micromhos	Mhos	10^6
Micromicrofarads	Attofarads	10^6
Micromicrofarads	Farads	10^{-12}
Micromicrofarads	Femtofarads	10^3
Micromicrofarads	Microfarads	10^{-6}
Micromicrofarads	Nanofarads	10^{-3}
Micromicrofarads	Picofarads	1
Micromicrons	Centimeters	10^{-10}
Micromicrons	Meters	10^{-12}
Micromicrons	Microns	10^{-6}
Microns	Centimeters	10^{-4}
Microns	Feet	3.281×10^{-6}
Microns	Inches	3.937×10^{-5}
Microns	Kilometers	10^{-9}
Microns	Megameters	10^{-12}
Microns	Meters	10^{-6}

| --- | --- | --- |
Microns	Millimeters	10^{-3}
Microns	Millimicrons	10^3
Microns	Mils	0.03937
Microns	Yards	1.09361×10^{-6}
Microroentgens	Kiloroentgens	10^{-9}
Microroentgens	Megaroentgens	10^{-12}
Microroentgens	Milliroentgens	10^3
Microroentgens	Roentgens	10^{-6}
Microrutherfords	Kilorutherfords	10^{-9}
Microrutherfords	Megarutherfords	10^{-12}
Microrutherfords	Millirutherfords	10^{-3}
Microrutherfords	Rutherfords	10^{-6}
Microseconds	Days	1.157×10^{-11}
Microseconds	Hours	2.77×10^{-10}
Microseconds	Milliseconds	10^{-3}
Microseconds	Minutes	1.667×10^{-8}
Microseconds	Nanoseconds	10^3
Microseconds	Picoseconds	10^6
Microseconds	Seconds	10^{-6}
Microvolts	Attovolts	10^{12}
Microvolts	Femtovolts	10^9
Microvolts	Gigavolts	10^{-15}
Microvolts	Kilovolts	10^{-9}
Microvolts	Megavolts	10^{-12}
Microvolts	Nanovolts	10^3
Microvolts	Millivolts	10^{-3}
Microvolts	Picovolts	10^6
Microvolts	Teravolts	10^{-18}
Microvolts	Volts	10^{-6}
Microvolts/meter	Millivolts/meter	10^{-3}
Microvolts/meter	Volts/meter	10^{-6}
Microwatts	Attowatts	10^{12}
Microwatts	Femtowatts	10^9
Microwatts	Gigawatts	10^{-15}
Microwatts	Horsepower	1.3×10^{-9}
Microwatts	Kilowatts	10^{-9}
Microwatts	Megawatts	10^{-12}
Microwatts	Milliwatts	10^{-3}
Microwatts	Nanowatts	10^3
Microwatts	Picowatts	10^6

To Convert From	To	Multiply By
Microwatts	Terawatts	10^{-18}
Microwatts	Watts	10^{-6}
Miles	Astronomical units	1.0764×10^{-9}
Miles	Centimeters	1.60935×10^5
Miles	Feet	5280
Miles	Inches	6.3360×10^4
Miles	Kilometers	1.60935
Miles	Light years	1.70122×10^{-13}
Miles	Meters	1609.35
Miles	Millimeters	1.609347×10^6
Miles	Myriameters	0.160935
Miles	Parsecs	5.208×10^{-14}
Miles	Yards	1760
Miles/hr	Centimeters/sec	44.7
Miles/hr	Feet/hr	5280
Miles/hr	Feet/min	88
Miles/hr	Feet/sec	1.4667
Miles/hr	Kilometers/hr	1.6093
Miles/hr	Kilometers/min	0.02682
Miles/hr	Meters/min	26.82
Miles/hr	Meters/sec	0.4470
Miles/hr	Miles/min	0.01667
Miles/hr	Miles/sec	2.78×10^{-4}
Miles/min	Centimeters/sec	2682.2
Miles/min	Feet/hr	3.168×10^5
Miles/min	Feet/min	5280
Miles/min	Feet/sec	88
Miles/min	Kilometers/hr	96.561
Miles/min	Kilometers/min	1.6093
Miles/min	Meters/min	1609.35
Miles/min	Meters/sec	26.82
Miles/min	Miles/hr	60
Miles/sec	Feet/hr	1.9×10^7
Miles/sec	Feet/sec	5280
Miles/sec	Feet/min	88
Miles/sec	Miles/hr	3600
Miles/sec	Miles/min	60
Milliampere-hours	Ampere-hours	10^{-3}
Milliampere-hours	Faradays	37.31
Milliamperes	Amperes	10^{-3}

To Convert From	To	Multiply By
Milliamperes	Attoamperes	10^{15}
Milliamperes	Coulombs/sec	10^{-3}
Milliamperes	Femtoamperes	10^{12}
Milliamperes	Kiloamperes	10^{-6}
Milliamperes	Megamperes	10^{-9}
Milliamperes	Microamperes	10^{3}
Milliamperes	Nanoamperes	10^{6}
Milliamperes	Picoamperes	10^{9}
Milliamperes/volt	Amperes/volt	10^{-3}
Milliamperes/volt	Microamperes/volt	10^{3}
Milliangstroms	Angstroms	10^{-3}
Milliangstroms	Microangstroms	10^{3}
Millibars	Bars	10^{-3}
Millibars	Dynes/sq cm	10^{3}
Millibars	Grams/sq cm	1.01972
Millibars	Microbars	10^{3}
Millibars	Millitorrs	7.5×10^{2}
Millibars	Torrs	0.750
Millicuries	Curies	10^{-3}
Millicuries	Kilocuries	10^{-6}
Millicuries	Megacuries	10^{-9}
Millicuries	Microcuries	10^{3}
Milligrams	Grams	10^{-3}
Milligrams	Kilograms	10^{-6}
Milligrams	Micrograms	10^{3}
Milligrams	Ounces (avdp)	3.5274×10^{-5}
Milligrams	Pounds (avdp)	2.2046×10^{-6}
Milligrams/cm	Dynes/cm	0.980665
Milligrams/cm	Dynes/in.	2.491
Milligrams/cm	Grams/cm	10^{-3}
Milligrams/cm	Grams/in.	2.54×10^{-3}
Milligrams/cm	Kilograms/km	0.1
Milligrams/cm	Pounds/ft	6.72×10^{-5}
Milligrams/cm	Pounds/in.	5.599×10^{-6}
Milligrams/in.	Dynes/cm	0.3861
Milligrams/in.	Dynes/in.	1.9807
Milligrams/in.	Grams/in.	3.937×10^{-4}
Milligrams/in.	Grams/in.	10^{-3}
Milligrams/in.	Kilograms/km	0.03937
Milligrams/in.	Pounds/ft	2.645×10^{-5}

To Convert From	To	Multiply By
Milligrams/in.	Pounds/in.	2.2046×10^{-6}
Milligrams/liter	Grams/cu cm	9.9997×10^{-7}
Milligrams/liter	Grams/liter	10^{-3}
Milligrams/liter	Pounds/cu ft	6.24×10^{-5}
Milligrams/mm	Dynes/cm	9.807
Milligrams/mm	Dynes/in.	24.909
Milligrams/mm	Grams/cm	10^{-2}
Milligrams/mm	Grams/in.	0.0254
Milligrams/mm	Kilograms/km	1
Milligrams/mm	Pounds/ft	6.71×10^{-4}
Milligrams/mm	Pounds/in.	5.52×10^{-5}
Millihenrys	Henrys	10^{-3}
Millihenrys	Microhenrys	10^{3}
Millihenrys	Nanohenrys	10^{6}
Millihenrys	Picohenrys	10^{9}
Millilamberts	Candles/sq cm	3.18×10^{-4}
Millilamberts	Candles/sq ft	0.2957
Millilamberts	Candles/sq in.	2.05×10^{-3}
Millilamberts	Lamberts	10^{-3}
Milliliters	Cubic cm	1
Milliliters	Cubic in.	0.06102
Milliliters	Gallons (liq)	2.64×10^{-4}
Milliliters	Gills	8.45×10^{-3}
Milliliters	Liters	10^{-3}
Milliliters	Microliters	10^{3}
Milliliters	Pints (liq)	2.11×10^{-3}
Milliliters	Quarts (liq)	1.05×10^{-3}
Millilix	Lux	10^{-3}
Millilux	Microlux	10^{3}
Millimaxwells	Maxwells	10^{-3}
Millimeters	Centimeters	0.1
Millimeters	Feet	3.281×10^{-3}
Millimeters	Inches	0.03937
Millimeters	Kilometers	10^{6}
Millimeters	Megameters	10^{-9}
Millimeters	Meters	10^{-3}
Millimeters	Micromicrons	10^{9}
Millimeters	Microns	10^{3}
Millimeters	Millimicrons	10^{6}
Millimeters	Miles	6.2137×10^{-7}

To Convert From	To	Multiply By
Millimeters	Mils	39.37
Millimeters	Yards	1.094×10^{-3}
Millimicrons	Centimeters	10^{-7}
Millimicrons	Feet	3.2808×10^{-9}
Millimicrons	Inches	3.937×10^{-8}
Millimicrons	Kilometers	10^{-12}
Millimicrons	Meters	10^{-9}
Millimicrons	Microns	10^{-3}
Millimicrons	Millimeters	10^{-6}
Millimicrons	Mils	3.937×10^{-5}
Millimicrons	Yards	1.0936×10^{-9}
Millimoles	Moles	10^{-3}
Milliohms	Gigohms	10^{-12}
Milliohms	Kilohms	10^{-6}
Milliohms	Megohms	10^{-9}
Milliohms	Microhms	10^3
Milliohms	Ohms	10^{-3}
Milliohms	Teraohms	10^{-15}
Milliohms/cm	Milliohms/in.	2.54
Milliohms/cm	Milliohms/mil	2.54×10^{-3}
Milliohms/in.	Milliohms/cm	0.3937
Milliohms/in.	Milliohms/mil	10^{-3}
Milliohms/mil	Milliohms/cm	393.7
Milliohms/mil	Milliohms/in.	10^3
Milliphots	Meter candles	10
Milliphots	Phots	10^{-3}
Milliroentgens	Kiloroentgens	10^{-6}
Milliroentgens	Megaroentgens	10^{-9}
Milliroentgens	Microroentgens	10^3
Milliroentgens	Roentgens	10^{-3}
Millirutherfords	Kilorutherfords	10^{-6}
Millirutherfords	Megarutherfords	10^{-9}
Millirutherfords	Microrutherfords	10^3
Millirutherfords	Rutherfords	10^{-3}
Milliseconds	Days	1.157×10^{-8}
Milliseconds	Hours	2.77×10^{-7}
Milliseconds	Microseconds	10^3
Milliseconds	Minutes	1.667×10^{-5}
Milliseconds	Nanoseconds	10^6
Milliseconds	Picoseconds	10^9

To Convert From	To	Multiply By
Milliseconds	Seconds	10^{-3}
Millitorrs	Bars	1.333×10^{-6}
Millitorrs	Microbars	13.33
Millitorrs	Millibars	1.333×10^{-3}
Millitorrs	Torrs	10^{-3}
Millivolts	Attovolts	10^{15}
Millivolts	Femtovolts	10^{12}
Millivolts	Gigavolts	10^{-12}
Millivolts	Kilovolts	10^{-6}
Millivolts	Megavolts	10^{-9}
Millivolts	Microvolts	10^{3}
Millivolts	Nanovolts	10^{6}
Millivolts	Picovolts	10^{9}
Millivolts	Teravolts	10^{-15}
Millivolts	Volts	10^{-3}
Millivolts/meter	Microvolts/meter	10^{3}
Millivolts/meter	Volts/meter	10^{-3}
Milliwatts	Attowatts	10^{15}
Milliwatts	Femtowatts	10^{12}
Milliwatts	Gigawatts	10^{-12}
Milliwatts	Horsepower	1.34×10^{-10}
Milliwatts	Kilowatts	10^{-6}
Milliwatts	Megawatts	10^{-9}
Milliwatts	Microwatts	10^{3}
Milliwatts	Nanowatts	10^{6}
Milliwatts	Picowatts	10^{9}
Milliwatts	Terawatts	10^{-15}
Milliwatts	Watts	10^{-3}
Milliwebers	Microwebers	10^{3}
Milliwebers	Webers	10^{-3}
Mils	Centimeters	2.54×10^{-3}
Mils	Feet	8.3333×10^{-5}
Mils	Inches	10^{-3}
Mils	Kilometers	2.54×10^{-8}
Mils	Microns	25.4
Mils	Millimeters	0.0254
Mils	Yards	2.7778×10^{-5}
Minutes (angle)	Circles	4.62×10^{-5}
Minutes (angle)	Circumferences	4.62×10^{-5}
Minutes (angle)	Degrees	1.667×10^{-2}

To Convert From	To	Multiply By
Minutes (angle)	Grads	1.852×10^{-2}
Minutes (angle)	Quadrants	1.852×10^{-3}
Minutes (angle)	Radians	2.91×10^{-4}
Minutes (angle)	Seconds	60
Minutes (time)	Days	6.944×10^{-4}
Minutes (time)	Hours	0.016667
Minutes (time)	Microseconds	6×10^{7}
Minutes (time)	Milliseconds	6×10^{4}
Minutes (time)	Nanoseconds	6×10^{10}
Minutes (time)	Picoseconds	6×10^{13}
Minutes (time)	Seconds	60
Moles	Millimoles	10^{3}
Myriagrams	Grams	10^{4}
Myriagrams	Kilograms	10
Myriagrams	Pounds (avdp)	22.05
Myriameters	Kilometers	10
Myriameters	Meters	10^{4}
Myriameters	Miles	6.214
Nanoamperes	Amperes	10^{-9}
Nanoamperes	Attoamperes	10^{9}
Nanoamperes	Coulombs/sec	10^{-9}
Nanoamperes	Femtoamperes	10^{6}
Nanoamperes	Kiloamperes	10^{-12}
Nanoamperes	Megamperes	10^{-15}
Nanoamperes	Microamperes	10^{-3}
Nanoamperes	Milliamperes	10^{-6}
Nanoamperes	Picoamperes	10^{3}
Nanofarads	Attofarads	10^{9}
Nanofarads	Farads	10^{-9}
Nanofarads	Femtofarads	10^{6}
Nanofarads	Microfarads	10^{-3}
Nanofarads	Micromicrofarads	10^{3}
Nanofarads	Picofarads	10^{3}
Nanohenrys	Henrys	10^{-9}
Nanohenrys	Microhenrys	10^{-3}
Nanohenrys	Millihenrys	10^{-6}
Nanohenrys	Picohenrys	10^{3}
Nanoseconds	Days	1.157×10^{-14}
Nanoseconds	Hours	2.77×10^{-13}
Nanoseconds	Microseconds	10^{-3}

To Convert From	To	Multiply By
Nanoseconds	Milliseconds	10^{-6}
Nanoseconds	Minutes	1.667×10^{-11}
Nanoseconds	Picoseconds	10^3
Nanoseconds	Seconds	10^{-9}
Nanovolts	Attovolts	10^9
Nanovolts	Femtovolts	10^6
Nanovolts	Gigavolts	10^{-18}
Nanovolts	Kilovolts	10^{-12}
Nanovolts	Megavolts	10^{-15}
Nanovolts	Microvolts	10^{-3}
Nanovolts	Millivolts	10^{-6}
Nanovolts	Picovolts	10^3
Nanovolts	Teravolts	10^{-21}
Nanovolts	Volts	10^{-9}
Nanowatts	Attowatts	10^9
Nanowatts	Femtowatts	10^6
Nanowatts	Gigawatts	10^{-18}
Nanowatts	Horsepower	1.34×10^{-12}
Nanowatts	Kilowatts	10^{-12}
Nanowatts	Megawatts	10^{-15}
Nanowatts	Microwatts	10^{-3}
Nanowatts	Milliwatts	10^{-6}
Nanowatts	Picowatts	10^3
Nanowatts	Terawatts	10^{-21}
Nanowatts	Watts	10^{-9}
Natural log	Common log	0.4343
Ohm-cm	Ohm-in.	0.3937
Ohm-in.	Ohm-cm	2.54
Ohms	Abohms	10^9
Ohms	Gigohms	10^{-9}
Ohms	Kilohms	10^{-3}
Ohms	Megohms	10^{-6}
Ohms	Microhms	10^6
Ohms	Milliohms	10^3
Ohms	Teraohms	10^{-12}
Ohms/cm	Ohms/in.	2.54
Ohms/cm	Ohms/mil	2.54×10^{-3}
Ohms/in.	Ohms/cm	0.3937
Ohms/in.	Ohms/mil	10^{-3}
Ohms/mil	Ohms/cm	393.7

To Convert From	To	Multiply By
Ohms/mil	Ohms/in.	10^3
Ohms/volt	Kilohms/volt	10^{-3}
Ohms/volt	Megohms/volt	10^{-6}
Ounces (avdp)	Grams	28.35
Ounces (avdp)	Hectograms	0.2835
Ounces (avdp)	Kilograms	0.02835
Ounces (avdp)	Milligrams	2.83495×10^4
Ounces/sq ft	Dynes/sq cm	29.925
Ounces/sq ft	Grams/sq cm	0.0305
Ounces/sq in.	Dynes/sq cm	4309.2
Ounces/sq in.	Dynes/sq cm	4309.2
Ounces/sq in.	Grams/sq cm	4.394
Parsecs	Astronomical units	4.7893×10^{-6}
Parsecs	Light years	3.26
Parsecs	Miles	1.92×10^{13}
Phots	Foot-candles	929.03
Phots	Lumens/sq cm	1
Phots	Lumens/sq ft	929.03
Phots	Lumens/sq meter	10^4
Phots	Lux	10^4
Phots	Milliphots	10^3
Picoamperes	Amperes	10^{-12}
Picoamperes	Attoamperes	10^6
Picoamperes	Coulombs/sec	10^{-12}
Picoamperes	Femtoamperes	10^3
Picoamperes	Kiloamperes	10^{-15}
Picoamperes	Megamperes	10^{-18}
Picoamperes	Microamperes	10^{-6}
Picoamperes	Milliamperes	10^{-9}
Picoamperes	Nanoamperes	10^{-3}
Picofarads	Attofarads	10^6
Picofarads	Farads	10^{-12}
Picofarads	Femtofarads	10^3
Picofarads	Microfarads	10^{-6}
Picofarads	Micromicrofarads	1
Picofarads	Nanofarads	10^{-3}
Picohenrys	Henrys	10^{-12}
Picohenrys	Microhenrys	10^{-6}
Picohenrys	Micromicrohenrys	1
Picohenrys	Millihenrys	10^{-9}

To Convert From	To	Multiply By
Picohenrys	Nanohenrys	10^{-3}
Picoseconds	Days	1.157×10^{-17}
Picoseconds	Hours	2.77×10^{-16}
Picoseconds	Microseconds	1×10^{-6}
Picoseconds	Milliseconds	1×10^{-9}
Picoseconds	Minutes	1.667×10^{-14}
Picoseconds	Nanoseconds	10^{-3}
Picoseconds	Seconds	10^{-12}
Picovolts	Attovolts	10^6
Picovolts	Femtovolts	10^3
Picovolts	Gigavolts	10^{-21}
Picovolts	Kilovolts	10^{-15}
Picovolts	Megavolts	10^{-18}
Picovolts	Microvolts	10^{-6}
Picovolts	Millivolts	10^{-9}
Picovolts	Nanovolts	10^{-3}
Picovolts	Teravolts	10^{-24}
Picovolts	Volts	10^{-12}
Picowatts	Attowatts	10^6
Picowatts	Femtowatts	10^3
Picowatts	Gigawatts	10^{-21}
Picowatts	Horsepower	1.31×10^{-15}
Picowatts	Kilowatts	10^{-15}
Picowatts	Megawatts	10^{-18}
Picowatts	Microwatts	10^{-6}
Picowatts	Milliwatts	10^{-9}
Picowatts	Nanowatts	10^{-3}
Picowatts	Terawatts	10^{-24}
Picowatts	Watts	10^{-12}
Pints (liq)	Cubic cm	473.2
Pints (liq)	Cu meters	4.732×10^{-4}
Pints (liq)	Gallons (liq)	0.125
Pints (liq)	Gills	4
Pints (liq)	Liters	0.4732
Pints (liq)	Milliliters	473.2
Pints (liq)	Quarts (liq)	0.5
Planck's constants	Erg-sec	6.6236×10^{-27}
Planck's constants	Joule-sec	6.6236×10^{-34}
Poises	Centipoises	100
Poundals	Dynes	1.3825×10^4

To Convert From	To	Multiply By
Poundals	Grams	14.098
Poundals	Kilograms	0.0141
Pounds (avdp)	Grams	453.59
Pounds (avdp)	Kilograms	0.4539
Pounds (avdp)	Milligrams	4.535924×10^5
Pounds/cu ft	Grams/cu cm	0.01602
Pounds/cu ft	Grams/liter	16.018
Pounds/cu ft	Kilograms/cu meter	16.018
Pounds/cu ft	Kilograms/hectoliter	1.6018
Pounds/cu in.	Grams/cu cm	27.679
Pounds/cu in.	Grams/liter	2.7679×10^4
Pounds/cu in.	Kilograms/cu meter	2.7679×10^4
Pounds/cu in.	Kilograms/hectoliter	2768.05
Pounds/ft	Grams/cm	14.882
Pounds/ft	Grams/ft	453.6
Pounds/ft	Grams/in.	37.799
Pounds/ft	Kilograms/foot	0.4536
Pounds/ft	Kilograms/kilometer	1488.2
Pounds/ft	Kilograms/meter	1.488
Pounds/ft	Ounces/cm	0.5249
Pounds/ft	Pounds/meter	3.2808
Pounds/in.	Grams/cm	178.58
Pounds/in.	Grams/ft	5443.11
Pounds/in.	Grams/in.	453.59
Pounds/in.	Kilograms/foot	5.4431
Pounds/in.	Kilograms/kilometer	1.7858×10^4
Pounds/in.	Kilograms/meter	17.858
Pounds/in.	Ounces/cm	6.299
Pounds/in.	Pounds/meter	39.37
Pounds/sq ft	Dynes/sq cm	478.8
Pounds/sq ft	Grams/sq cm	0.4882
Pounds/sq ft	Kilograms/sq cm	4.882×10^{-4}
Pounds/sq ft	Kilograms/sq meter	4.882
Pounds/sq in.	Dynes/sq cm	6.89476×10^4
Pounds/sq in.	Grams/sq cm	70.307
Pounds/sq in.	Kilograms/sq cm	0.07031
Pounds/sq in.	Kilograms/sq meter	703.067
Quadrants	Circles	0.25
Quadrants	Circumferences	0.25
Quadrants	Degrees	90

To Convert From	To	Multiply By
Quadrants	Grads	100
Quadrants	Minutes	5400
Quadrants	Radians	1.57
Quadrants	Revolutions	0.25
Quadrants	Seconds	3.24×10^5
Quarts (liq)	Cubic cm	946.36
Quarts (liq)	Cubic meters	9.46×10^{-4}
Quarts (liq)	Gallons	0.25
Quarts (liq)	Gills	8
Quarts (liq)	Liters	0.946
Quarts (liq)	Milliliters	946.3
Quarts (liq)	Pints	2
Quintals (metric)	Grams	10^5
Quintals (metric)	Kilograms	10^2
Quintals (metric)	Pounds (avdp)	220.46
Radians	Circles	0.159
Radians	Circumferences	0.159
Radians	Degrees	57.2958
Radians	Grads	63.654
Radians	Minutes	3437.75
Radians	Quadrants	0.636
Radians	Revolutions	0.1591
Radians	Right angles	0.6366
Radians	Seconds	2.06265×10^5
Radians/sec	Degrees/sec	57.296
Radians/sec	Grads/sec	63.66
Revolutions	Circumferences	1
Revolutions	Degrees	360
Revolutions	Quadrants	4
Revolutions	Radians	6.283
Revolutions/min	Degrees/sec	6
Revolutions/min	Grads/sec	6.6667
Revolutions/min	Radians/sec	0.10472
Revolutions/min	Revolutions/sec	0.016667
Revolutions/sec	Degrees/sec	360
Revolutions/sec	Grads/sec	400
Revolutions/sec	Radians/sec	6.283
Right angles	Degrees	90
Right angles	Grads	100
Right angles	Radians	1.571

To Convert From	To	Multiply By
RMS value (sine wave)	Average value	0.899
RMS value (sine wave)	Effective value	1
RMS value (sine wave)	Maximum value	1.414
RMS value (square wave)	Average value	1
RMS value (square wave)	Effective value	1
RMS value (square wave)	Maximum value	1
Roentgens	Kiloroentgens	10^{-3}
Roentgens	Megaroentgens	10^{-6}
Roentgens	Microroentgens	10^6
Roentgens	Milliroentgens	10^3
Rutherfords	Curies	2.69×10^{-5}
Rutherfords	Kilorutherfords	10^{-3}
Rutherfords	Megarutherfords	10^{-6}
Rutherfords	Microrutherfords	10^6
Rutherfords	Millirutherfords	10^3
Seconds (angle)	Circumferences	7.716×10^{-4}
Seconds (angle)	Degrees	2.78×10^{-4}
Seconds (angle)	Grads	3.086×10^{-4}
Seconds (angle)	Minutes	0.01667
Seconds (angle)	Quadrants	3.1×10^{-6}
Seconds (angle)	Radians	4.848×10^{-6}
Seconds (time)	Days	1.15×10^{-5}
Seconds (time)	Hours	2.777×10^{-4}
Seconds (time)	Microseconds	10^6
Seconds (time)	Milliseconds	10^3
Seconds (time)	Minutes	0.016667
Seconds (time)	Nanoseconds	10^9
Seconds (time)	Picoseconds	1×10^{12}
Square cm	Acres	2.471×10^{-8}
Square cm	Ares	10^{-6}
Square cm	Circ in.	0.1973
Square cm	Circ mm	127.32
Square cm	Circ mils	1.97352×10^5
Square cm	Hectares	10^{-8}

To Convert From	To	Multiply By
Square cm	Sq decimeters	10^{-2}
Square cm	Sq feet	1.076×10^{-3}
Square cm	Sq inches	0.15499
Square cm	Sq kilometers	10^{-10}
Square cm	Sq meters	10^{-4}
Square cm	Sq miles	3.86×10^{-11}
Square cm	Sq mm	100
Square cm	Sq mil	1.55×10^5
Square cm	Sq yards	1.19×10^{-4}
Square cm/dyne	Sq cm/gram	980.665
Square cm/dyne	Sq cm/kilogram	9.80665×10^5
Square cm/dyne	Sq ft/pound	478.8
Square cm/dyne	Sq in./pound	6.8947×10^4
Square cm/gram	Sq cm/dyne	1.02×10^{-3}
Square cm/kg	Sq cm/dyne	1.0197×10^{-6}
Square decameters	Acres	0.0247
Square decameters	Ares	1
Square decameters	Hectares	10^{-2}
Square decameters	Sq centimeters	10^6
Square decameters	Sq feet	1076.4
Square decameters	Sq inches	1.55×10^5
Square decameters	Sq meters	100
Square decameters	Sq yards	119.59
Square decimeters	Sq centimeters	100
Square decimeters	Sq feet	0.1076
Square decimeters	Sq inches	15.5
Square decimeters	Sq kilometers	10^{-8}
Square decimeters	Sq meters	10^{-2}
Square decimeters	Sq millimeters	10^4
Square feet	Acres	2.2957×10^{-5}
Square feet	Ares	9.29×10^{-4}
Square feet	Hectares	9.29×10^{-6}
Square feet	Sq centimeters	929.03
Square feet	Sq kilometers	9.29×10^{-8}
Square feet	Sq meters	0.0929
Square feet	Sq millimeters	9.2903×10^4
Square hectometers	Acres	2.47
Square hectometers	Ares	10^2
Square hectometers	Centares	10^4
Square hectometers	Hectares	1

To Convert From	To	Multiply By
Square hectometers	Sq centimeters	10^8
Square hectometers	Sq feet	1.07639×10^5
Square hectometers	Sq inches	1.55×10^4
Square hectometers	Sq kilometers	10^{-2}
Square hectometers	Sq meters	10^4
Square hectometers	Sq miles	3.86×10^{-3}
Square hectometers	Sq yards	1.19598×10^4
Square inches	Acres	1.594×10^{-7}
Square inches	Ares	6.451×10^{-6}
Square inches	Hectares	6.451×10^{-8}
Square inches	Sq centimeters	6.451
Square inches	Sq decimeters	0.06451
Square inches	Sq kilometers	6.451×10^{-10}
Square inches	Sq meters	6.451×10^{-4}
Square inches	Sq millimeters	645.16
Square kilometers	Acres	247.1
Square kilometers	Ares	10^4
Square kilometers	Hectares	10^2
Square kilometers	Sq centimeters	10^{10}
Square kilometers	Sq feet	1.076×10^7
Square kilometers	Sq inches	1.55×10^9
Square kilometers	Sq meters	10^6
Square kilometers	Sq miles	0.3861
Square kilometers	Sq millimeters	10^{12}
Square kilometers	Sq yards	1.196×10^6
Square meters	Acres	2.47×10^{-4}
Square meters	Ares	10^{-2}
Square meters	Centares	1
Square meters	Hectares	10^{-4}
Square meters	Sq centimeters	10^4
Square meters	Sq decameters	10^{-2}
Square meters	Sq feet	10.764
Square meters	Sq inches	1549.9
Square meters	Sq kilometers	10^{-6}
Square meters	Sq miles	3.861×10^{-7}
Square meters	Sq millimeters	10^6
Square meters	Sq yards	1.1959
Square miles	Acres	640
Square miles	Ares	2.5899×10^4
Square miles	Hectares	258.9

To Convert From	To	Multiply By
Square miles	Sq centimeters	2.589×10^{10}
Square miles	Sq decimeters	2.589×10^{8}
Square miles	Sq kilometers	2.589
Square miles	Sq meters	2589
Square miles	Sq millimeters	2.589×10^{12}
Square millimeters	Circ miles	1973.52
Square millimeters	Sq centimeters	10^{-2}
Square millimeters	Sq feet	1.076×10^{-5}
Square millimeters	Sq inches	1.549×10^{-3}
Square millimeters	Sq kilometers	10^{-12}
Square millimeters	Sq meters	10^{-6}
Square millimeters	Sq miles	3.861×10^{-13}
Square millimeters	Sq mils	1550
Square millimeters	Sq yards	1.1959×10^{-6}
Square mils	Sq centimeters	6.452×10^{-6}
Square mils	Sq millimeters	6.452×10^{-4}
Square yards	Acres	2.066×10^{-4}
Square yards	Ares	8.361×10^{-3}
Square yards	Hectares	8.361×10^{-5}
Square yards	Sq centimeters	8361.3
Square yards	Sq decameters	8.361×10^{-3}
Square yards	Sq kilometers	8.3613×10^{-7}
Square yards	Sq meters	0.83613
Square yards	Sq millimeters	8.3613×10^{5}
Statamperes	Abamperes	3.336×10^{-11}
Statamperes	Amperes (abs)	3.336×10^{-10}
Statcoulombs	Abcoulombs	3.336×10^{-11}
Statcoulombs	Ampere-hours	9.266×10^{-4}
Statcoulombs	Coulombs (abs)	3.336×10^{-10}
Statcoulombs	Faradays	3.457×10^{-15}
Statcoulombs/dyne	Abcoulombs/kg	3.271×10^{-5}
Statcoulombs/dyne	Abcoulombs/lb	1.484×10^{-5}
Statcoulombs/dyne	Coulombs (abs)/kg	3.271×10^{-4}
Statcoulombs/dyne	Faradays/kg	3.39×10^{-9}
Statcoulombs/dyne	Statcoulombs/kg	9.807×10^{5}
Statcoulombs/dyne	Statcoulombs/lb	4.48×10^{5}
Statcoulombs/kg	Statcoulombs/dyne	1.0197×10^{-6}
Statcoulombs/lb	Statcoulombs/dyne	2.248×10^{-6}
Statcoulombs/sq cm	Abcoulombs/sq cm	3.336×10^{-11}
Statcoulombs/sq cm	Coulombs (abs)/sq cm	3.336×10^{-10}

To Convert From	To	Multiply By
Statcoulombs/sq cm	Coulombs (abs)/sq in.	2.152×10^{-9}
Statfarads	Abfarads	1.113×10^{-21}
Statfarads	Farads (abs)	1.113×10^{-12}
Statfarads	Microfarads (abs)	1.113×10^{-6}
Stathenrys	Abhenrys	8.985×10^{20}
Stathenrys	Henrys (abs)	8.9864×10^{11}
Stathenrys	Microhenrys (abs)	8.9864×10^{17}
Stathenrys	Millihenrys (abs)	8.9864×10^{14}
Statmhos	Mhos (abs)	1.1128×10^{-12}
Statohms	Abohms	8.9864×10^{20}
Statohms	Megohms (abs)	8.9864×10^{5}
Statohms	Microhms (abs)	8.9864×10^{17}
Statohms	Milliohms (abs)	8.9864×10^{14}
Statohms	Ohms (abs)	8.8964×10^{11}
Statvolts	Abvolts	2.997×10^{10}
Statvolts	Microvolts (abs)	2.997×10^{8}
Statvolts	Millivolts (abs)	2.9977×10^{5}
Statvolts	Volts (abs)	299.77
Statwebers	Maxwells (abs)	2.9977×10^{10}
Statwebers	Webers	299.77
Stokes	Centistokes	10^{2}
Steres	Cu meters	1
Steres	Decasteres	0.1
Steres	Decisteres	10
Steres	Liters	10^{3}
Teracycles	Cycles	10^{12}
Teracycles	Gigacycles	10^{3}
Teracycles	Kilocycles	10^{9}
Teracycles	Megacycles	10^{6}
Teracycles/sec	Cycles/sec	10^{12}
Teracycles/sec	Gigacycles/sec	10^{3}
Teracycles/sec	Gigahertz	10^{3}
Teracycles/sec	Hertz	10^{12}
Teracycles/sec	Kilocycles/sec	10^{9}
Teracycles/sec	Kilohertz	10^{9}
Teracycles/sec	Megacycles/sec	10^{6}
Teracycles/sec	Megahertz	10^{6}
Teracycles/sec	Terahertz	1
Terahertz	Cycles/sec	10^{12}
Terahertz	Gigacycles/sec	10^{3}

To Convert From	To	Multiply By
Terahertz	Gigahertz	10^3
Terahertz	Hertz	10^{12}
Terahertz	Kilocycles/sec	10^9
Terahertz	Kilohertz	10^9
Terahertz	Megacycles/sec	10^6
Terahertz	Megahertz	10^6
Terahertz	Teracycles/sec	1
Teraohms	Gigohms	10^3
Teraohms	Kilohms	10^9
Teraohms	Megohms	10^6
Teraohms	Microhms	10^{18}
Teraohms	Milliohms	10^{15}
Teraohms	Ohms	10^{12}
Teraohms/cm	Teraohms/in.	2.54
Teraohms/cm	Teraohms/mil	0.00254
Teraohms/in.	Teraohms/cm	0.3937
Teraohms/in.	Teraohms/mil	10^{-3}
Teraohms/mil	Teraohms/cm	393.7
Teraohms/mil	Teraohms/in.	10^3
Teravolts	Attovolts	10^{30}
Teravolts	Femtovolts	10^{27}
Teravolts	Gigavolts	10^3
Teravolts	Kilovolts	10^9
Teravolts	Megavolts	10^6
Teravolts	Microvolts	10^{18}
Teravolts	Millivolts	10^{15}
Teravolts	Nanovolts	10^{21}
Teravolts	Picovolts	10^{24}
Teravolts	Volts	10^{12}
Terawatts	Attowatts	10^{30}
Terawatts	Femtowatts	10^{27}
Terawatts	Gigawatts	10^3
Terawatts	Horsepower	1.34×10^9
Terawatts	Kilowatts	10^9
Terawatts	Megawatts	10^6
Terawatts	Microwatts	10^{18}
Terawatts	Milliwatts	10^{15}
Terawatts	Nanowatts	10^{21}
Terawatts	Picowatts	10^{24}
Terawatts	Watts	10^{12}

To Convert From	To	Multiply By
Tons (long)	Grams	1.016×10^6
Tons (long)	Kilograms	1016
Tons (long)	Pounds (avdp)	2240
Tons (long)	Tons (metric)	1.016
Tons (long)	Tons (short)	1.12
Tons (metric)	Grams	10^6
Tons (metric)	Kilograms	10^3
Tons (metric)	Pounds (avdp)	2204.6
Tons (metric)	Tons (long)	0.9842
Tons (metric)	Tons (short)	1.1023
Tons (short)	Grams	9.07185×10^5
Tons (short)	Kilograms	907.185
Tons (short)	Pounds (avdp)	2000
Tons (short)	Tons (long)	0.8928
Tons (short)	Tons (metric)	0.9072
Torrs	Bars	1.333×10^{-3}
Torrs	Microbars	1.333×10^3
Torrs	Millibars	1.333
Torrs	Millitorrs	10^3
Turns/cm	Turns/in.	2.54
Turns/in.	Turns/cm	0.3937
Volt-amperes	Attowatts	10^{18}
Volt-amperes	Femtowatts	10^{15}
Volt-amperes	Gigawatts	10^{-9}
Volt-amperes	Kilovolt-amperes	10^{-3}
Volt-amperes	Kilowatts	10^{-3}
Volt-amperes	Megavolt-amperes	10^{-6}
Volt-amperes	Megawatts	10^{-6}
Volt-amperes	Microwatts	10^6
Volt-amperes	Milliwatts	10^3
Volt-amperes	Nanowatts	10^9
Volt-amperes	Picowatts	10^{12}
Volt-amperes	Terawatts	10^{-12}
Volts	Abvolts	10^8
Volts	Attovolts	10^{18}
Volts	Femtovolts	10^{15}
Volts	Gigavolts	10^{-9}
Volts	Kilovolts	10^{-3}
Volts	Megavolts	10^{-6}
Volts	Microvolts	10^6

To Convert From	To	Multiply By
Volts	Millivolts	10^3
Volts	Nanovolts	10^{-9}
Volts	Picovolts	10^{12}
Volts	Statvolts	3.33×10^{-3}
Volts	Teravolts	10^{-12}
Volts/cm	Attovolts/cm	10^{18}
Volts/cm	Femtovolts/cm	10^{15}
Volts/cm	Gigavolts/cm	10^{-9}
Volts/cm	Kilovolts/cm	10^{-3}
Volts/cm	Megavolts/cm	10^{-6}
Volts/cm	Microvolts/cm	10^6
Volts/cm	Microvolts/meter	10^8
Volts/cm	Millivolts/cm	10^3
Volts/cm	Millivolts/meter	10^5
Volts/cm	Nanovolts/cm	10^9
Volts/cm	Picovolts/cm	10^{12}
Volts/cm	Teravolts/cm	10^{-12}
Volts/cm	Volts/in.	2.54
Volts/cm	Volt/mil	2.54×10^{-3}
Volts/cm	Volts/mm	0.1
Volts/in.	Attovolts/in.	10^{18}
Volts/in.	Femtovolts/in.	10^{15}
Volts/in.	Gigavolts/in.	10^{-9}
Volts/in.	Kilovolts/cm	3.937×10^{-4}
Volts/in.	Kilovolts/in.	10^{-3}
Volts/in.	Megavolts/in.	10^{-6}
Volts/in.	Microvolts/in.	10^6
Volts/in.	Microvolts/meter	3.937×10^{-7}
Volts/in.	Millivolts/in.	10^3
Volts/in	Millivolts/meter	3.9370×10^4
Volts/in.	Nanovolts/in.	10^9
Volts/in.	Picovolts/in.	10^{12}
Volts/in.	Teravolts/in.	10^{-12}
Volts/in.	Volts/cm	0.3937
Volts/in.	Volts/mil	10^{-3}
Volts/in.	Volts/mm	0.03937
Volts/meter	Microvolts/meter	10^6
Volts/meter	Millivolts/meter	10^3
Volts/mil	Attovolts/mil	10^{18}
Volts/mil	Femtovolts/mil	10^{15}

To Convert From	To	Multiply By
Volts/mil	Gigavolts/mil	10^{-9}
Volts/mil	Kilovolts/cm	0.3937
Volts/mil	Kilovolts/mil	10^{-3}
Volts/mil	Megavolts/mil	10^{-6}
Volts/mil	Microvolts/meter	3.937×10^{10}
Volts/mil	Microvolts/mil	10^6
Volts/mil	Millivolts/meter	3.937×10^7
Volts/mil	Millivolts/mil	10^3
Volts/mil	Nanovolts/mil	10^9
Volts/mil	Picovolts/mil	10^{12}
Volts/mil	Teravolts/mil	10^{-12}
Volts/mil	Volts/cm	393.7
Volts/mil	Volts/in.	10^3
Volts/mil	Volts/mm	39.37
Volts/mm	Kilovolts/cm	10^{-2}
Volts/mm	Microvolts/meter	10^9
Volts/mm	Millivolts/meter	10^6
Volts/mm	Volts/cm	10
Volts/mm	Volts/in.	25.4
Volts/mm	Volts/mil	0.0254
Volts/sq cm	Volts/sq in.	0.155
Volts/sq in.	Volts/sq cm	6.452
Watt-hours	Foot-pounds	2655.7
Watt-hours	Horsepower-hr	1.34×10^{-3}
Watt-hours	Joules	3600
Watt-hours	Kilogram-meters	367.2
Watt-hours	Kilowatt-hours	10^{-3}
Watt-hours	Watt-seconds	3000
Watts	Attowatts	10^{18}
Watts	Ergs/sec	10^7
Watts	Femtowatts	10^{15}
Watts	Foot-poundals/min	1424.1
Watts	Foot-pounds/min	44.26
Watts	Foot-pounds/sec	0.738
Watts	Gigawatts	10^{-9}
Watts	Gram-cm/sec	1.0199×10^4
Watts	Hectowatts	10^{-2}
Watts	Horsepower	1.34×10^{-3}
Watts	Joules/sec	1

To Convert From	To	Multiply By
Watts	Kilogram-meters/sec	0.1019
Watts	Kilowatts	10^{-3}
Watts	Lumens	680
Watts	Megawatts	10^{-6}
Watts	Microwatts	10^6
Watts	Milliwatts	10^3
Watts	Nanowatts	10^9
Watts	Picowatts	10^{12}
Watts	Terawatts	10^{-12}
Watt-seconds	Foot-pounds	0.7377
Watt-seconds	Horsepower-hr	3.726×10^{-7}
Watt-seconds	Joules	1
Watt-seconds	Kilogram-meters	0.1019
Watt-seconds	Kilowatt-hours	2.778×10^{-7}
Watt-seconds	Watt-hours	2.778×10^{-4}
Webers	Kilolines	10^5
Webers	Lines	10^8
Webers	Maxwells	10^8
Webers	Microwebers	10^6
Webers	Milliwebers	10^3
Webers/sq cm	Gausses	10^8
Webers/sq cm	Kilolines/sq cm	10^5
Webers/sq cm	Lines/sq cm	10^8
Webers/sq cm	Lines/sq in.	6.4516×10^8
Webers/sq cm	Maxwells/sq cm	10^8
Webers/sq cm	Webers/sq in.	6.4516
Webers/sq in.	Gausses	1.55×10^7
Webers/sq in.	Lines/sq cm	1.5×10^7
Webers/sq in.	Lines/sq in.	10^8
Webers/sq in.	Maxwells/sq cm	1.55×10^7
Webers/sq in.	Maxwells/sq in.	10^8
Webers/sq in.	Webers/sq cm	0.155
Yards	Centimeters	91.44
Yards	Feet	3
Yards	Hectometers	9.144×10^{-3}
Yards	Inches	36
Yards	Kilometers	9.144×10^{-4}
Yards	Meters	0.9144
Yards	Microns	914.4

To Convert From	To	Multiply By
Yards	Miles	5.682×10^{-4}
Yards	Millimeters	914.4
Yards	Millimicrons	9.144×10^{8}
Yards	Mils	3.6×10^{4}

Chapter 16

Schematic Symbols

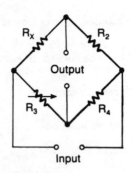

AMMETER	
AND GATE	
ANTENNA, BALANCED	
ANTENNA, GENERAL	
ANTENNA, LOOP, SHIELDED	
ANTENNA, LOOP, UNSHIELDED	
ANTENNA, UNBALANCED	
ATTENUATOR, FIXED	
ATTENUATOR, VARIABLE	
BATTERY	
CAPACITOR, FEEDTHROUGH	
CAPACITOR, FIXED, NONPOLARIZED	
CAPACITOR, FIXED, POLARIZED	
CAPACITOR, GANGED, VARIABLE	

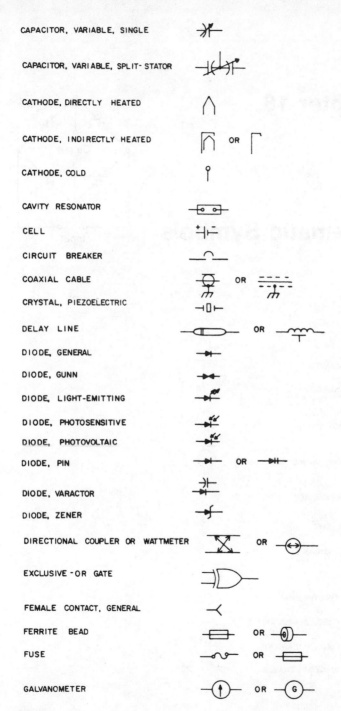

CAPACITOR, VARIABLE, SINGLE	
CAPACITOR, VARIABLE, SPLIT-STATOR	
CATHODE, DIRECTLY HEATED	
CATHODE, INDIRECTLY HEATED	OR
CATHODE, COLD	
CAVITY RESONATOR	
CELL	
CIRCUIT BREAKER	
COAXIAL CABLE	OR
CRYSTAL, PIEZOELECTRIC	
DELAY LINE	OR
DIODE, GENERAL	
DIODE, GUNN	
DIODE, LIGHT-EMITTING	
DIODE, PHOTOSENSITIVE	
DIODE, PHOTOVOLTAIC	
DIODE, PIN	OR
DIODE, VARACTOR	
DIODE, ZENER	
DIRECTIONAL COUPLER OR WATTMETER	OR
EXCLUSIVE-OR GATE	
FEMALE CONTACT, GENERAL	
FERRITE BEAD	OR
FUSE	OR
GALVANOMETER	OR

252

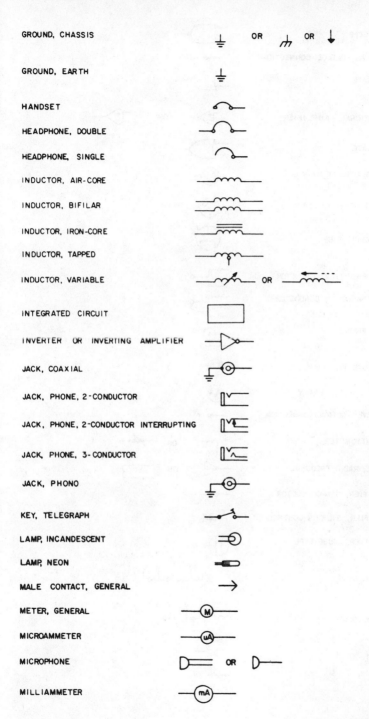

GROUND, CHASSIS

GROUND, EARTH

HANDSET

HEADPHONE, DOUBLE

HEADPHONE, SINGLE

INDUCTOR, AIR-CORE

INDUCTOR, BIFILAR

INDUCTOR, IRON-CORE

INDUCTOR, TAPPED

INDUCTOR, VARIABLE

INTEGRATED CIRCUIT

INVERTER OR INVERTING AMPLIFIER

JACK, COAXIAL

JACK, PHONE, 2-CONDUCTOR

JACK, PHONE, 2-CONDUCTOR INTERRUPTING

JACK, PHONE, 3-CONDUCTOR

JACK, PHONO

KEY, TELEGRAPH

LAMP, INCANDESCENT

LAMP, NEON

MALE CONTACT, GENERAL

METER, GENERAL

MICROAMMETER

MICROPHONE

MILLIAMMETER

253

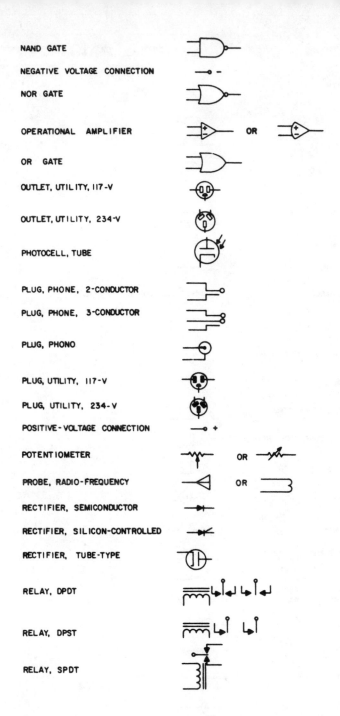

NAND GATE

NEGATIVE VOLTAGE CONNECTION

NOR GATE

OPERATIONAL AMPLIFIER   OR

OR GATE

OUTLET, UTILITY, 117-V

OUTLET, UTILITY, 234-V

PHOTOCELL, TUBE

PLUG, PHONE, 2-CONDUCTOR

PLUG, PHONE, 3-CONDUCTOR

PLUG, PHONO

PLUG, UTILITY, 117-V

PLUG, UTILITY, 234-V

POSITIVE-VOLTAGE CONNECTION

POTENTIOMETER   OR

PROBE, RADIO-FREQUENCY   OR

RECTIFIER, SEMICONDUCTOR

RECTIFIER, SILICON-CONTROLLED

RECTIFIER, TUBE-TYPE

RELAY, DPDT

RELAY, DPST

RELAY, SPDT

254

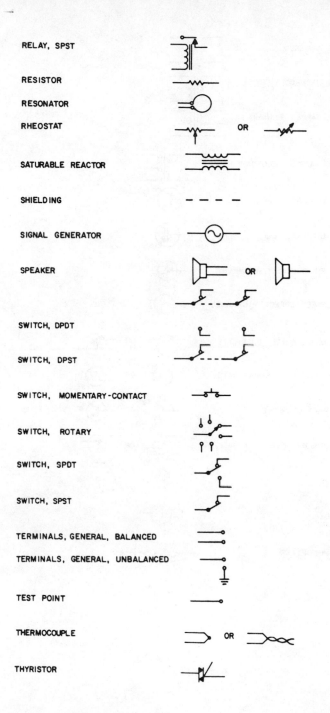

RELAY, SPST

RESISTOR

RESONATOR

RHEOSTAT   OR

SATURABLE REACTOR

SHIELDING

SIGNAL GENERATOR

SPEAKER   OR

SWITCH, DPDT

SWITCH, DPST

SWITCH, MOMENTARY-CONTACT

SWITCH, ROTARY

SWITCH, SPDT

SWITCH, SPST

TERMINALS, GENERAL, BALANCED

TERMINALS, GENERAL, UNBALANCED

TEST POINT

THERMOCOUPLE   OR

THYRISTOR

255

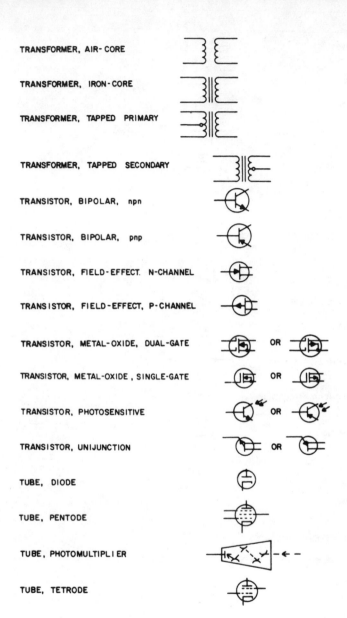

TRANSFORMER, AIR-CORE

TRANSFORMER, IRON-CORE

TRANSFORMER, TAPPED PRIMARY

TRANSFORMER, TAPPED SECONDARY

TRANSISTOR, BIPOLAR, npn

TRANSISTOR, BIPOLAR, pnp

TRANSISTOR, FIELD-EFFECT. N-CHANNEL

TRANSISTOR, FIELD-EFFECT, P-CHANNEL

TRANSISTOR, METAL-OXIDE, DUAL-GATE   OR

TRANSISTOR, METAL-OXIDE, SINGLE-GATE   OR

TRANSISTOR, PHOTOSENSITIVE   OR

TRANSISTOR, UNIJUNCTION   OR

TUBE, DIODE

TUBE, PENTODE

TUBE, PHOTOMULTIPLIER

TUBE, TETRODE

Notes

Notes

Notes

Notes

Notes

Notes

Notes

Notes

Notes

Notes

Index

Other Bestsellers From TAB